Helion & Company Limited
Unit 8 Amherst Business Centre
Budbrooke Road
Warwick
CV34 5WE
England
Tel. 01926 499 619
Email: info@helion.co.uk
Website: www.helion.co.uk
Twitter: @helionbooks
Visit our blog http://blog.helion.co.uk/

Designed & typeset by Farr out Publications, Wokingham, Berkshire
Cover design Paul Hewitt, Battlefield Design (www.battlefield-design.co.uk)
Printed by Henry Ling Limited, Dorchester, Dorset

ISBN 978-1-913118-69-3

British Library Cataloguing-in-Publication Data
A catalogue record for this book is available from the British Library

We always welcome receiving book proposals from prospective authors.

CONTENTS

Note: in order to simplify the use of this book, all names, locations and geographic designations are as provided in *The Times World Atlas*, or other traditionally accepted major sources of reference, as of the time of the described events. All military ranks are as in official use at the time of the described events. Unless stated otherwise, all times are local times. While only the first letter of the official designations of institutions and services of the Yugoslav armed forces was written in capital letters, and this practice was subsequently adopted in Croatia and Slovenia, their translations in this book are written as usual in the English language. When mentioned for the first time in the text, aircraft and heavy weapon system designations are cited fully – including their designer and/or the manufacturer, official military designation and nickname: in the case of Soviet-made armament: this is followed by the ASCC/NATO-codename, but through the text only their original designations are used. Local, Yugoslav designations are provided in separate tables. Unless cited in the endnotes, all sources are interviews with participants or original documentation listed in the bibliography.

ABBREVIATIONS

AAM	air to air missile
AB	air base
AGM	air to ground missile
AFAC	airborne forward air controller
CAP	combat air patrol
CAS	close air support
ЦК ПУАК	Центральные ордена Красной Звезды Курсы по подготовке и усовершенствованию авиационных кадров (Order of Red Star Central Course for Training and Improvement of Aviation Personnel, Soviet Air Force)
CO	commanding officer
ELINT	electronic intelligence
HAS	hardened aircraft shelter
HDA	Hrvatski Drzavni arhiv (Croatian State Archive)
FRY	Federal Republic of Yugoslavia
IFR	instrumental flying rules
IOC	initial operational capability
JAT	Jugoslovenski Aero Transport (Yugoslav Air Transport)
HQ	headquarters
L	Lovac ('fighter' in Serbo-Croatian)
LORAP	long range aerial panoramic photography
MANPADS	man-portable air defence system
MDAP	Mutual Defence Aid Programme
MM	mlazni motor ('jet engine', in Serbo-Croatian)
NATO	North Atlantic Treaty Organization
NCO	non-commissioned officer
OCU	operational conversion unit
PVO	protivvazdušna odbrana ('air defence', in Serbo-Croatian)
QRA	quick reaction alert
RV i PVO	Ratno vayduhoplovstvo i protivvazdušna odbrana ('Air Force and Air Defence Force', in Serbo-Croatian)
SAM	surface-to-air missile
SFRJ	Socijalistička Federativna Republika Jugoslavia (Socialist Federative Republic of Yugoslavia)
SIGINT	signals intelligence
TDY	temporary duty
UN	United Nations
UNPROFOR	United Nations Protection Force
USAF	United States Air Force
VA	Vojni arhiv (Military Archive, Belgrade)
VJ	Vojska Jugoslavije (Armed Forces of the FRY)
VS	Vazduhoplovstvo Srbije (Serbian Air Force)
VFR	visual flying rules

INTRODUCTION

The Soviet-made Mikoyan i Gurevich (MiG) MiG-21 jet fighter was truly a legend of the Cold War. Over the last 60 years it has served with more than 50 air forces around the world – more than any other fighter type before or after. For many services in the countries that were members of the Warsaw Pact, and in many so-called Third World countries, it was the first supersonic – and at that the first Mach 2-capable – fighter jet, and its primary role was that of a fighter-interceptor. Nevertheless, it was often deployed as a multi-role fighter, and as such saw combat operations in dozens of wars in Africa, Asia, Europe, and the Middle East. Unsurprisingly, its impressive historical account has been published in a vast number of books, and this volume thus might appear as about to join the long line of similar publications. However, this operational history of MiG-21s in service with the Air Force and Air Defence Force (Ratno vazduhoplovstvo i protivvazdušna odbrana, RV i PVO) of Yugoslavia, from 1962 until 1992, the RV I PVO of the Yugoslav Army (Vojska Jugoslavije, VJ) from 1992-2006, and then the Serbian Air Force (Vazduhoplovstvo Srbije, VS), from 2006 until the present, is significantly different to anything published on this topic before. It is based on extensive research in archives and museums, and upon interviews with retired and active participants and eyewitnesses, and thus provides a particularly authoritative and detailed insight into the numerous crises of the Cold War and the four bitter wars in the former Socialist Federative Republic of Yugoslavia (Socijalistička Federativna Republika Jugoslavia, SFRJ), from 1991 until 1999.

The book is written in chronological order and is organised in seven chapters describing the MiG-21's service from the first deliveries, through the following two decades and on to the combat deployments in Slovenia, Croatia, Bosnia and Herzegovina, the 'shortened' Yugoslavia, and then in Serbia in the early years of the 21st century. As such, it offers not only the operational history of this type, but also that of the Yugoslav and then Serbian Air Force *de-facto* until the time of publication.

Table 1: RV i PVO's Designations for MiG-21 Aircraft

Original Designation	RV i PVO Designation
MiG-21F-13	L-12
MiG-21U-400/600	NL-12
MiG-21PFM	L-14
MiG-21R	L-14R later L-14i
MiG-21US	NL-14
MiG-21M	L-15
MiG-21MF	L-15M
MiG-21MF (LORAP mod.)	L-15i
MiG-21UM	NL-16
MiG-21bis	L-17
MiG-21bis-K	L-17K
MiG-21bis (Iraqi aircraft)	L-17IQ

1
PLAN GROM

During the 1950s, Yugoslavia maintained relatively close relations to the United States of America (USA). Its representatives signed the Mutual Defence Aid Program (MDAP) and the armed forces thus received significant amounts of arms, equipment, and weapon systems of Western origin. However, after the end of the MDAP in 1958, the government in Beograd (Belgrade) – capital of the SFRJ – turned to commercial deliveries for equipping its air force. Following contemporary trends, the RV i PVO designed a plan for its further growth for 1959-1966, which included an estimated requirement for 80 fighter jets capable of reaching speeds of Mach 2. It might sound surprising for what at the time was a communist country, but the senior leadership of the RV i PVO was quick to conclude that its preferred type would be the French-made Dasault Mirage IIIC interceptor; indeed, the generals were even interested in obtaining a licence for its manufacture in Yugoslavia.[1] However, at the same time, the Liberation War of Algeria was in full swing, and the government in Belgrade was providing political support to the National Liberation Front (FLN) in that country: indeed, it went as far as to recognise its provisional government in exile, and started providing arms deliveries. It is hardly necessary to say that this not only hampered relations with Paris but led to a cancellation of the Mirage order.

Instead, and just as the Cold War was about to enter the period of its next major crisis, the Yugoslavs turned to the Union of the Soviet Socialist Republics (USSR or Soviet Union). A visit of the Yugoslav Air Force commander General Ulepič in 1960 opened the possibility of Yugoslavia purchasing new fighters in the Soviet Union. A trade agreement between the two states for the period between 1961 and 1966 included an agreement on weapon deliveries including 34 million US dollars devoted to purchases for the air force. The primary Yugoslav aim was to obtain a modern supersonic fighter-interceptor. Although having only very few hard facts about the type – mostly obtained from various newspapers – the senior leadership of the RV i PVO decided that the MiG-21 would be its choice. On 17 April 1961, General Ulepič approved a plan for purchasing the MiG-21 and the licence for its production, codenamed Thunder ('Grom').[2]

Plan Grom estimated that the purchase and licence production would include 150 MiG-21s as well as the purchase of 12 MiG-15UTI two-seaters for training. It is interesting that the Yugoslav military authorities, despite knowing next to nothing about the new type, decided to launch large-scale licence production of the MiG-21 at the Soko Factory in Mostar, in Bosnia and Herzegovina, even though this lacked any kind of experience in manufacturing jet aircraft. Indeed, they went as far as to consider licence production of the Tumansky R-11F-300 engine for the new type. A Yugoslav delegation reached Moscow on 19 April 1961 and remained in the Soviet Union until August. On 22 April, it was given the opportunity to see a demonstration flight of a MiG-21F-13 at Kubinka Air Base, and then received a – rather basic – briefing on its performance.[3] Moreover, during the following negotiations, the Soviet representatives demanded that the Yugoslavs 'significantly change their political course' from Belgrade's pro-West orientation towards closer relations with the Warsaw Pact.[4]

Surprisingly enough, considering their fierce insistence on independence, the politicians in Belgrade agreed with such demands. Indeed, they advised Ulepič to abandon the earlier practice of getting closely acquainted with the performance of the new jet and pressed on

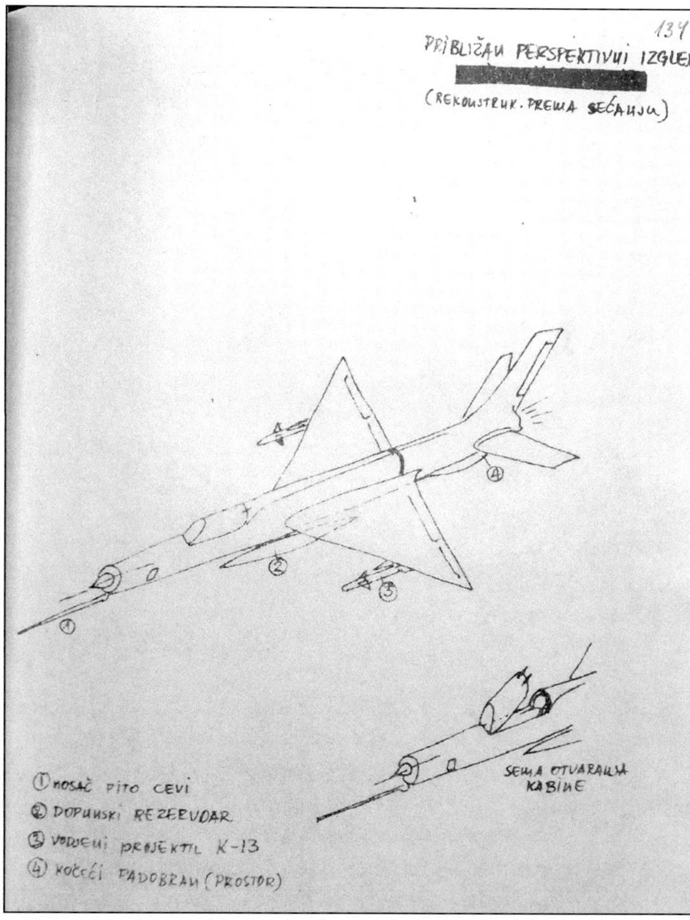

A sketch drawing made after the first demonstration of the MiG-21F-13 to the Yugoslav delegation. The Yugoslavs were not allowed to photograph the aircraft, so one of the colonels, Ing Nebojša Zivanović, drew this sketch for official use. (VA)

with negotiating the order. Correspondingly, and although the Soviets refused to lower the price, on 8 June 1961 the Yugoslav delegation announced its decision to purchase a batch of MiG-21F-13s. The only element of the original intention of the RV i PVO left out from the resulting contract was the acquisition of MiG-15UTI conversion trainers. Once all the related details were sorted out, on 2 August 1961 representatives of the SFRJ and the USSR signed an interstate agreement (соглашение), without details of quantities and prices, in Moscow. This stipulated the delivery of the first five MiG-21F-13s in 1962, eight in 1963, 14 in 1964, and 13 in 1965.[5]

In March 1962, a Soviet delegation visited Belgrade to negotiate a decision to purchase 40 MiG-21F-13s worth US$ 575,000, including spare parts, R-3S infra-red homing air-to-air missiles (ASCC/NATO-code 'AA-2 Atoll'; colloquially known by its weapons system designation 'R-3S' in the RV i PVO), ground equipment, and the training of necessary personnel of the RV i PVO. A related deal was signed on 7 April 1962 which specified the purchase of 40 MIG-21Fs.[6]

First Conversion of the Yugoslav Pilots

In January 1962, a group of RV i PVO airmen was selected for a trip to the USSR to convert to MiG-21 fighters. There were seven pilots, 20 technicians and one doctor. This group went through a two-month intensive Russian language course at the Air Technical Centre in

The Mikoyan-Gurevich MiG-21F-13 or L-12 as it was called in Yugoslavia. It was the Yugoslav choice to replace ageing US-supplied jets. (Author's collection)

СВИДЕТЕЛЬСТВО № 2/18

Настоящее свидетельство выдано:

Ончевски Ангел

(фамилия, имя и отчество)

в том, что он в 1962 г. поступил и в 1962 г.

окончил курсы по специальности (специализации)

летчик истребительной авиации

При Центральных Курсах по подготовке и усовершенствованию авиационных кадров

(наименование военно-учебного заведения)

Все требования, предъявленные программами и учебным планом выполнил.

Начальник ЦК ПУАК генерал-майор авиации

15 августа 1962 г.

Приложение к свидетельству № 2/18

(без свидетельства недействительно)

ОНЧЕВСКИ АНГЕЛ

(фамилия, имя и отчество)

за время обучения показал следующую успеваемость по предметам:

1. Теория полета -отлично
2. Авиатехника -отлично
3. Р Т С -отлично
4. В С П -отлично
5. Парашютная подготовка-отлично
6.
7.
8.
9.
10.
11.
12.
13.
14.

Soviet diploma issued to the Yugoslav pilots after the conversion course during 1962. (Angel Ončevski)

Rajlovac, near Sarajevo between 28 February and 25 April. Nobody knew what kind of the task the group was destined for except the group's leader. The group was led by pilot Major Slobodan Rakić, seasoned F-86E Sabre pilot who finished conversion on the F-100 Super Sabre in USAF in 1957-1958, and the technicians were led by Captain 1st Class Ivan Dubravčić. On May Day 1962, this group boarded a train for the Soviet Union. After reaching Moscow, they continued to Frunze to the 5th Order of Red Star Central Course for Training and Improvement of Aviation Personnel (5 Центральные ордена Красной Звезды Курсы по подготовке и усовершенствованию авиационных кадров, 5 ЦК ПУАК), also known as the Frunze Military Aviation School for Pilots (Фрунзенское Военное Авиационное Училище Летчиков). Although all of the pilots passed strict medical exams in Yugoslavia, at the request of the Soviets they were examined again prior to starting the conversion course and one of the Yugoslav pilots was returned by the Soviets. The Soviet style of training was from the ground-up, and treated the Yugoslav pilots as though they had never flown fighters before, no matter their previous experience on US-built jets. Pilots were converted to fly the MiG-21 in daytime/VFR conditions by 715 Training Aviation Regiment (715 учебный авиационный полк) at Lugovaya AB in Kirgizia. They conducted ten sorties on MiG-15UTIs and a dozen on MiG-17s between 1 June and 4 July. From 4 July, they started to fly MiG-21s until the middle of August, reaching an average of 23 training sorties per pilot. The first to make a Mach-2 sonic boom was Major Slobodan Rakić, on 20 July, followed by the other pilots. After the Yugoslav pilots learned how to take off, make a circle around the air base and land, the Soviets provided no tactical training whatsoever. Ground crew proceed from Lugovaya to Alma Ata where they joined the technicians from other air forces which were also about to receive the MiG-21s, and passed the training to maintain the MiGs at 'I and II level', i.e. at the squadron and regimental levels.[7]

Conversion was finished and on 20 August the group returned to Yugoslavia with the technicians proudly stating that they were the best among other foreign groups to finish conversion. Thus came into being the nucleus of the Yugoslav supersonic fighter community.

An excellent study of MiG-21F-13 No. 22510, seen at Batajnica in 1963. This particular MiG would be lost in a fatal crash during Marshal Tito's visit to Bihać Air Base on 3 May 1969. (Author's collection)

The rear view of MiG-21F-13 No. 22502 at Batajnica in 1963. It was lost in a fatal crash on 18 June 1970. (Author's collection)

Initial Deliveries

On 22 September 1962, the first five MiG-21F-13s arrived at Batajnica Air Base, some 15 miles north of Belgrade.[8] According to the contract with the Soviets, the Yugoslav RV i PVO increased security measures at Batajnica Air Base to prevent any leakage of information concerning this type. The sales contract also requested secrecy over the type's name and so Yugoslavia adopted their own type designations; the MiG-21F-13 became the L-12, where "L" stood for *lovac* or fighter, while the later-delivered MiG-21U became the NL-12 where "NL" meant *nastavni lovac*, or training-fighter. This pattern developed as further versions were delivered and remained throughout the whole of the MiG-21's service. Nearly all of the Yugoslav air force personnel (as well as the authors of this volume) used this system to mark specific MiG-21 versions. This system sometimes did not follow the differences in the original subversions or series, which may have been referred to as the same "L". On the other hand, local modifications were always added into the broader "L" title.

The first batch of five MiG-21F-13s were produced by Gorkovski Aviatsioni Zavod which ceased to produce this type by the end of 1962. The export production of F-13s continued in the Znamja Truda factory in Moscow. The first five MiGs did not have hardpoints for carrying bombs, and the first 13 also lacked the SRO-2 IFF device, though much later, in 1966, those airframes were fitted with the missing equipment by Soviet technicians.[9] Since the MiG-21F-13 version ceased production in favour of the new PF version, the last batch of 13 airframes was delivered in 1964 instead in 1965.[10] Over the following two years the delivery of the MiG-21F-13 was completed totalling 40 airframes.

MiG-21Fs on the Batajnica flight-line in 1966. Note that the first three MiGs in the line, belonging to the first two batches delivered, were fitted with the missing SRO-2 IFF, and standardised with others that were delivered later. (Author's collection)

No. 22903 was one of the MiG-21Us (NL-12s) belonging to the batch which was delivered in July 1965 and was of the early series marked by Soviets as 66-400. It was preserved in Zadar – Zemunik Air Base. (MC Odbrana)

The first Yugoslav MiG-21 was written off shortly after on 20 June 1964, killing pilot Stanic onboard. Following an accident during take-off, on 8 April 1965, Major Ončevski managed to make an emergency landing and then safely bail out. An official investigation into the latter case run by regimental engineer, Captain Dubravčič, concluded that there was a mistake in the calibration of the fuel system, and that this was wrongly calibrated when fitted at the factory. The Soviet aviation authorities confirmed the Yugoslav conclusion and in 1966 provided a replacement MiG-21F-13.[11]

The first four two-seater MiG-21Us (NL-12s) were delivered to the RV i PVO in the summer of 1965, one year before scheduled.[12] These were early U/U-400 versions (Izdeliye-66-400) with the same narrow vertical stabiliser as the F-13 version. The other 14 were of a later version (Izdeliye-66-600) with a wider vertical stabiliser and breaking parachute.

Dressed in VKK-1 high altitude equipment with a Gsh-4M helmet, Major Ončevski of 204th Fighter Regiment in Batajnica poses near MiG-21F-13 No. 22501. (Angel Ončevski)

Conversion in 204th Fighter Regiment at Batajnica

The first unit to receive the Soviet-built fighters was 204th Fighter Aviation Regiment. Originally established in 1949, this regiment was based at Batajnica AB, west of Belgrade, from 1951. It operated MDAP-supplied Republic F-84G Thunderjets in 1953-1958 and then North American F-86E Sabres from 1958. When the MiGs arrived, the 204th consisted of two fighter squadrons (lovačka avijacijska eskadrila): Nos. 127 and 128, and the force's acrobatic team, and had around 25 F-86Es in total.

No. 127 Squadron was chosen to be the first unit to convert to MiG-21s; conversion started immediately after the MiGs arrived and continued until the end of 1963.[13] The pilots that finished conversion in the USSR continued flying in Yugoslavia from November 1962. In spring of 1963 conversion of the first group of five pilots started in Yugoslavia.[14] It is interesting to note that in 1962-1964, 204th Regiment was most likely unique in Europe and probably worldwide, to jointly operate American and Soviet fighters – and that at the peak of the Cold War!

The conversion programme continued in Batajnica in 1963. It included five more pilots and a group of technicians. Members of the first group which passed the conversion in the Soviet Union assisted in the process. Another group of 17 pilots passed the conversion between December 1963 and September 1964, followed by another group of 18 pilots from autumn 1964 to spring 1965. It took until spring 1965 for all technical personnel to be converted to maintain the type in so-called I and II degree of technical maintenance. In the meantime, there were also Soviet instructors in the regiment. A total of 54 different Soviet airmen were in the unit until 1966, assisting the conversion and inspecting the MiGs during the guarantee period. The Yugoslavs managed to overcome the Soviet limitations on flying the MiG-21F and to carry out all of the fighter training and even the acrobatics on their own.[15]

Conversion and initial training were carried out under VFR conditions, while flying under IFR rules was allowed only to the pilots of the group who had passed their conversion in the Soviet Union. Lack of flying gear also meant that most interceptions were carried at the middle altitudes and below Mach 1. Squadron exercises were also conducted: fast take offs from QRA, interception at middle and higher altitudes carried out by a single pilot or in a pair.

Compared to the earlier conversion to the American types delivered through the MDAP, the conversion to the Soviet supersonic fighter was burdened with many problems. The Soviet flying gear was in deficit. No official flying manuals, schemes, procedures or other devices used in preparation for flying were supplied. All of this was especially obvious when flying at higher altitudes and training for the sonic boom; it was not easy to go over the speed of sound just on the experiences of the elder pilots.

A major problem was the lack of the two-seater. The Soviets and their Warsaw Pact allies used the MiG-15UTI as the trainer for the MiG-21s until receipt of the production batches of two-seater MiG-21Us which did not arrive until 1965. Due to the lack of two-seaters during conversion, the Yugoslavs modified at least one Lockheed TV-2 (serial number 10258) in May 1964 by mounting Soviet metric instruments (air horizon) and continued conversion of other pilots using the modified American training aircraft: this was a unique application, but also a typical 'Yugoslav improvisation' – which

"Flying day" with No. 127 Squadron at Batajnica in 1963 reveals the first batches of the MiG-21F-13 delivered to the Yugoslav RV i PVO. (Author's collection)

Another "flying day" in Batajnica, most likely in 1964. The pilots are dressed in a variety of Yugoslav, US and Soviet flying equipment and are being briefed by 204th Regiment CO, Lieutenant Colonel Slobodan Rakić. (Author's collection)

Two MiG-21Fs seen on the runway at Batajnica, during the May Day Parade in 1964. It was their first public appearance in Yugoslavia. (Author's collection)

worked flawlessly until the delivery of dedicated two-seat conversion trainers.[16] Indeed, the delivery of the first batch of two-seaters caused a major surprise, because it turned out that the instructor – who always occupied the rear seat – could see almost nothing at all during the take-off and landing. Nevertheless, new commander of RV I PVO, General Bubanj complained to the Soviets about this 'solution'.[17]

The MiG-21 made its public debut in Yugoslavia only during the May Day parade in Belgrade of 1964. At the time, nine of the jets and nine pilots were available, from 13 delivered aircraft and 11 airworthy pilots, and they all made a flypast.[18] Further deliveries in 1964 enabled 204th Fighter Regiment to convert to this type fully, even to expand its assets into three fighter squadrons, Nos. 126, 127, and – following the reorganisation of 8 May 1964 – No. 128.[19]

In 1965, one of the MiG-21 squadrons moved to Pleso-Zagreb Air Base where it took part in Exercise Lovćen, together with F-86D Sabre Dogs from 117th Regiment. However, and overall, the conversion to MiGs progressed rather slowly. Indeed, during the same year a study by the RV i PVO concluded that, 'MiG-pilots are not capable of interceptor-operations at speeds above Mach 1.3'. Furthermore, it observed that – while a modern fighter 'safe for flying' – the MiG-21F-13 had a 'short range… (and) minimal engine resources', and that the engine was sensitive to service use.[20]

In January 1963, a three-month training course for R-3S air-to-air missiles was carried out with the Soviet instructors. The PPP-13 preparation station for this type of missile was delivered in this period too, while the first batch of operational missiles arrived in June 1963. By the end of the same year, Belgrade signed an order for additional missiles and the final batches of R-3S and RM-3 target simulators are known to have been delivered in 1965.[21] The first live firing exercise was held in July 1965, at a range over the southern Adriatic Sea.[22]

Further Conversions and Improvement of Combat Capability

In May 1965, another Yugoslav military delegation was dispatched to the Soviet Union to arrange further deliveries in 1966-1970 for the RV i PVO. They were tasked to obtain a much more modern fighter

MiGs of 204th Regiment seen during an exercise at Pleso-Zagreb Air Base in 1965. (MC Odbrana)

Flightline at Batajnica featuring three of the NL-12 two-seaters which were mostly used by 128th Fighter Squadron for conversion from the late 1960s and throughout the 1970s. (Author's collection)

No. 22541 was delivered in May 1966 as a replacement for MiG-21F No. 22516 lost in a crash caused by an error at the factory where it was built. Seen here in Batajnica with all the fighter and strike ordnance it could carry: R-3S AAMs on under-wing pylons, NR-30 gun, FAB bombs and UB-16-57 unguided rocket launchers. On the left side is a Tumansky R-11F-300 engine. (Author's collection)

and reconnaissance version of the MiG-21 and integrated system of air surveillance and guidance of interceptors. On 3 June, the MiG-21PFM was shown to a delegation at Kubinka Air Base and four days later new types of radars and guidance system Vazduh 1P were demonstrated at the Air Defence Training Centre at Vladimir. The Yugoslavs were surprised with the complexity of the system and its price. An agreement was made for MiG-21PFM version and parts of the Vazduh 1P system due to the price and lack of the trained personnel. The Soviets explained that they could not deliver the reconnaissance version of the MiG-21R since it was not yet produced even for their air force.[23]

In February 1966 a group of three senior pilots were sent to Krasnodar Higher Military Aviation School of Pilots (Краснодарское высшее военное авиационное училище летчиков) in the USSR to convert to the MiG-21F and PFM and train in their tactical usage even though the PFM was not yet delivered. Furthermore, five instructors were also sent to be trained as instructors on MiG-21s. All of the groups were trained with 802nd Training Aviation Regiment (802-й учебный авиационный полк) at Krasnodar. Upon finishing the course in April of the same year, five instructors were sent to 128th Squadron.[24]

This third squadron of 204th Regiment was then converted to be the training and conversion unit for the MiG-21 and it would retain this status for a further 15 years, with all of the conversions conducted in 128th Squadron. This unit would receive nine of the two-seaters up to February 1967 and retain the larger portion of two-seaters inside its structure.[25] In this period, the squadron enabled the conversion on MiG-21F/L-12 type in VFR conditions. Advanced flying elements, flying in IFR conditions and combat training were carried out in the other squadrons.

In June 1966, the commander of 204th regiment, Lt Colonel Rakić, and Major Ončevski (both from the first group which converted in the USSR in 1962) were dispatched to 92nd Fighter Aviation Regiment (92-й истребительный

Краснознаменный авиационный полк) in Mukachevo, Ukraine, where they spent a month observing the operations of a regular Soviet fighter regiment. It seems that the Soviets were very receptive and showed the two Yugoslav commanders almost everything related to the functioning of their fighter unit and even let them take part in one of the regimental exercises. After their return, Rakić and Ončevski introduced their experiences to the functioning of 204th Fighter Regiment.[26]

The first group of Yugoslav airmen went to the Astrahan range in the USSR for live shooting in September 1966. They conducted live gunnery launching practice using R-3S AAMs within 116 Centre for Combat Aviation of the Soviet Air Defence Force (116-й Центр боевого применения авиации ПВО) based at Privolzhski AB. There were three senior officers including General Lekić, six pilots, and a team of armourers. The Yugoslavs used Soviet MiG-21F-13s and were assisted by their ground crews while preparing for training sorties during which they fired missiles at Lavochkin La-17 target drones and RM-3 target simulators.[27]

1966 was the first year where 204th Regiment reached serious combat capabilities. No. 126 Squadron carried out a course of IFR flying and practiced deployment to Tuzla Air Base. The other squadrons made deployments to Ladjevci – Kraljevo and Skopski Petrovac air bases. All three squadrons were able to carry out interceptions in daytime VFR conditions at speeds below Mach 1 at altitudes ranging from 9,000 to 12,000 metres, depending on the squadron. The pilots of No. 126 Squadron and commanders in No. 127 Squadron were converted to carry out interceptions at speeds beyond Mach 1 at altitudes up to 12,000 metres. The average annual flying hours per pilot

Dressed in flying suits, two young pilots are shown near the MiG-21Fs which were delivered in November 1964. Both of the MiGs successfully served within 204th, 117th and 83rd Fighter Regiments until 1981. (Author's collection)

The next step in MiG-21 usage in the Yugoslav RV i PVO was the PFM or L-14 version. Here No. 22709 demonstrates its appearance carrying only the underbelly 490-litre tank, known as "Bak". (Suad Hamzić)

Newly introduced to the RV i PVO inventory was the K-51 (RS-2US) radar-guided AAM. (Miroslav Špica)

were around 40, while the maximum was close to 60 flying hours per year. All of the squadrons were capable of carrying out strike missions, and aerial gunnery with 30mm guns, in which they were particularly successful. Serviceability was maintained at 80% but usually half of these airframes were used for daily flying. On the other hand, the MiG-21 fleet suffered from a weak capability of the air surveillance to guide interceptions successfully, and lack of sufficient and experienced MiG technicians and specific support vehicles, such as auxiliary power units (APA) which had a surprisingly high price for the Yugoslavs.[28]

A New Version Enters the Inventory: PFM or L-14

In 1967 further deliveries of the new MiG-21 versions continued and in the summer of 1967 a batch of new MiG-21PFMs were purchased. The new type featured many improvements compared to the earlier F-13 (L-12) version including an improved R-11F2S-300 engine, ARL-S LAZUR system for guidance during interception, R-3S and K-51 AAMs, GP-9 gun-pod with GSh-23L gun, SRZO-2M IFF transponder and above all the RP-21M radar which gave the definitive shape to the fuselage for all MiG-21s to come.

Before the delivery, a group of six pilots and 15 technicians, mostly from No. 126 Squadron, were sent to Krasnodar in the USSR for conversion to the type. Conversion at 802-й учебный авиационный полк lasted from February to June 1967 and pilots were converted for flying in VFR and IFR conditions.[29]

A total of 36 L-14s arrived at Batajnica in four batches between 21 June 1967 and 23 December 1968 where they started to re-equip No. 126 Squadron of 204th Regiment. The conversion was backed by the presence of two Soviet fighter pilots who demonstrated the new elements such as taking off with afterburner or from a grass strip.[30] This introduction enabled the withdrawal of the MiG-21F-13 (L-12) from the regiment's inventory and these were sent to Pleso Air Base near Zagreb for the conversion of 117th Fighter Aviation Regiment. Conversion of Nos. 127 and 128 squadrons continued and finished on the PFM in 1968. At the end of the conversion 44 pilots of 204th Regiment were capable of flying night missions on both L-12 and L-14 types. 204th Regiment continued training in interception at the whole range of altitudes in VFR and "above the clouds" during 1968. Interception tasks were carried out using "free hunt" patrols (i.e. with limited ground control), while the most experienced group was capable of dog fighting. During the following year a further group of 18 pilots carried out live firing over the south Adriatic with R-3S AAMs and the three most experienced pilots conducted live firing over the Astrahan range in the USSR, with MiG-21PFMs using the R-3S and, for the first time, with the RS-2US (ASCC/NATO-codename AA-1 Alkali) line-to-sight guided air-to-air missiles against La-17 drones. In 1969, 204th Fighter Regiment carried out live practice on the same variant with RS-2US AAMs at a range near the town of Bar, over the south Adriatic range, using the M-6 parachute target.[31]

Introduction of the new MiG-21 version improved the capabilities of the RV i PVO in general. Courses of instrumental flying were held on both versions. The young second lieutenants that arrived from the Air Academy in 1965-66 started conversion on MiGs in 1967. Moreover, the presence of the Soviet instructors was used to adopt new elements of flying techniques and combat usage such as group flying at low level, aerial combat/dogfighting and complex strike manoeuvres. It enabled 204th Regiment to create an informal demonstration team, and though this never appeared in public it showed its capabilities in front of several foreign military delegations in the following two years including the Soviets, East Germans and Hungarians, as well as a six-ship detachment tasked with carrying out air strikes with usage of complex manoeuvres.[32]

Three seasoned pilots were trained in the spring of 1968 at Krasnodar for interception at all altitudes and speeds, day or night, with active and passive jamming and usage of the KASKAD/LAZUR automated guidance system. (Momčilo Milenović)

MiG-21PFM, No. 22735, of 117th Fighter Regiment, seen in "fighter configuration" with two R-3S AAMs and a GP-9 gun-pod beneath the fuselage. (Mirko Popović)

Initially the PFMs were delivered without guns, and the first 18 delivered examples even did not have any installations for the cannons, though the other 18 did have.[33] Later, in the 1970s, a total of 39 GP-9 underbelly containers with GSh-23L guns were obtained in the Soviet Union to improve the strike capabilities of the L-14 community.[34]

Table 2: MiG-21s of the RV i PVO, 1 December 1967

Fighter Regiment	Squadron	Available aircraft and versions
204	HQ Det 126. 127. 128.	3 MiG-21PFM, 1 MiG-21U 15 MiG-21PFM, 2 MiG-21U 11 MiG-21F-13, 2 MiG-21U 11 MiG-21F-13, 11 MiG-21U
117	125.	12 MiG-21F-13, 2 MiG-21U
Zmaj Air Depot		4 MiG-21F-13
Total		18 MiG-21PFM, 38 MiG-21F-13, 18 MiG-21U

Enter 117th Regiment

The origins of 117th Fighter Aviation Regiment of the RV i PVO can be traced back to the final stages of the Second World War, when 168th Interceptor Aviation Regiment of the Soviet Air Force was temporarily deployed in Yugoslavia, to provide support for Partisan ground forces as these were fighting the Nazi Germans and their allies. The unit was transferred to the Yugoslav Army in May 1945, becoming 112th Fighter Regiment, and from the summer of 1948 it was renumbered to 117th Fighter Regiment. In 1949, this regiment was re-deployed to Belgrade and received the task of protecting the

Yugoslav capital at the time of a major showdown between the SFRJ and the Soviet-dominated Eastern Bloc. The regiment thus received an informal 'guards' status, and became the first to receive many fighter types, including the piston-engined Ikarus S-49s in 1950; MDAP-supplied T-33As in March 1953, and F-84Gs in June 1953; and then the F-86Es in May 1956. However, during the huge reorganisation of the RV i PVO, 117th Fighter Aviation Regiment was re-deployed to Pleso International Airport (IAP), outside Zagreb, and subsequently lost its priority within the air force in favour of 204th Regiment. Thus, it was not the first unit to convert to MiG-21s and instead in 1962-1963 it was re-equipped with second-hand North American F-86D Sabre Dogs. These jets were now in poor condition, lacking spares, radars, and even Mighty Mouse unguided air-to-air rockets, and the unit thus barely reached operational capability.[35] Nevertheless, as of 1964-1965, it was still the only interceptor unit of the RV i PVO capable of running intercept operations in all weather and at night.[36] Therefore, 117th Regiment continued flying Sabre Dogs until 1968.[37]

The conversion of 117th Regiment to MiGs began in 1967, when the first batch of MiG-21F-13s and MiG-21Us was provided to 21 pilots of its No. 125 Squadron and 17 of these successfully completed their course, which consisted of theoretical training at Pleso in May and June, followed by flight conversion at Batajnica in August and September of the same year. A month later, pilots that had completed their conversion transferred the first MiG-21F-13s to Pleso IAP.[38] The ground crew were also converted in this period, and proved 'satisfied' with the new type: contrary to the Sabre Dog, it contained no 'complicated electronics', there were enough spare parts, and refuelling was much easier. Meanwhile, the other squadron of this regiment – No. 124 – continued flying F-86Ds (indeed, it even borrowed a few obsolete F-86Es for re-training and conversion of older pilots) and standing the quick reaction alert (QRA). Thus, and once again, a regiment of the RV i PVO happened to operate both American and Soviet equipment at the same time, and that at the height of the Cold War!

In 1968, 117th Fighter Aviation Regiment was reinforced through the addition of the 352nd Reconnaissance Aviation Squadron. This unit came into being following the disbandment of 184th Reconnaissance Aviation Regiment, which used to operate North American IF-86Ds – a Yugoslav modification for reconnaissance purposes – in January 1966. The squadron retired its Sabre Dogs in 1968, and sent its personnel – led by Major Kulić and captains Kajević, Ficko and Stipić – to Krasondar, in the USSR, to convert to a new, dedicated reconnaissance variant of the MiG-21: the MiG-21R.[39]

In 1968, conversion of Nos. 124 and 125 Squadrons to MiG-21F-13 (locally designated the L-12) and the MiG-21U-400 (NL-12) was continued. As now usual, their personnel were also sent to Batajnica AB. Eventually, the regiment

Four pilots from 352nd Reconnaissance Squadron were trained in Krasnodar on reconnaissance MiG-21Rs in the first half of 1968. Here, 1st Class Captain Franc Ficko poses with a Soviet technician near a MiG which carries the "D" aerial photo reconnaissance pod beneath the fuselage. (Ficko family)

From May 1968, 117th Fighter Regiment became a permanent resident at the new air base near Bihać. Its core was underground Objekt Klek where three squadrons could be kept. Here, one of the MiG-21Fs is being towed into the underground objekt. (Author's collection)

MiG-21F-13 No. 22517 of 117th Fighter Regiment is seen in one of the galleries of the underground Objekt Klek in 1969. (HDA)

reached initial operational capability (IOC) when No. 125 Squadron completed its conversion at Pleso, in spring of 1968. Only weeks later, all elements of 117th Fighter Aviation Regiment received the order to move to the newly-constructed Bihać Air Base, and its Objekt Klek – a huge underground complex including kilometres-long galleries, workshops, communication and control facilities, and protection for pilots and ground crews.[40]

The first flying element of 117th to move to Bihać was the HQ element of No. 125 Fighter Aviation Squadron: it did so on personal order of the Commander RV i PVO, General Viktor Bubanj, to the newly-appointed regimental commander, Lieutenant-Colonel Ilija Zlatić (best known as one of the pilots that flew the yellow-overall-painted Sabres of the air force's aerobatic team in 1959-1963). Zlatić was the first to land a MiG-21F-13 on the runway of the newly-constructed Bihać AB – in heavy rain – on 27 May 1968.[41]

August 1968: Czechoslovak Crisis or Exercise Avala

When the aggression against Czechoslovakia was launched by the Soviets and Warsaw Pact in late August 1968 Tito's Yugoslavia felt a real danger for its independence. No matter its proclaimed non-alignment policy and siding with the Soviets since 1962, which was especially evident in the purchases of the Yugoslav Air Force, the threat was serious. Even more, the aggression against Czechoslovakia came as a total surprise to the Yugoslavs. Tito and his military and political leadership decided to raise the Yugoslav armed forces to the highest level of combat readiness and to deploy them to the state borders with the Warsaw Pact nations. On the night of 20-21 August, the whole air defence system was alerted. The whole operation was titled as Exercise Avala on 26 August 1968.[42]

204th Fighter Regiment with its three squadrons, as part of 11th Air Defence Division, was the tip of the spear of the Yugoslav air defence. 204th Regiment was ordered to disperse its squadrons: No. 127 remained in Batajnica, while No. 126 moved to Tuzla, and No. 128 to Ladjevci near Kraljevo. This last one was tasked to establish a fighter defence against any eventual threats from Bulgaria. It was a necessary step, since at that moment in Skoplje, No. 123 Fighter Squadron was finalising its conversion from F-86E Sabre to F-86D Sabre Dogs. The Regiment quickly established the QRA at the air bases where the squadrons were located and started to carry out the CAP missions in the eastern part of Yugoslavia. It was the first time that the regiment had operated at three different bases and this was a heavy burden on the regimental HQ. After eight days, No. 128 Squadron was ordered to move back to Batajnica. No. 126 Squadron remained in Tuzla until November, some time after the crisis calmed down.

During the time of the crisis, the alert was given using the signal 'Pirate' (*Gusar*). This was the signal that enemy formations were heading over Hungarian airspace towards Yugoslavia. The CO of the regiment, Lt. Colonel Savnik ordered all available MiGs and pilots, no matter their level of training, to take off immediately. It was a massive take off from Batajnica, but when the MiGs reached altitude the Operations Centre of 11th Division PVO cancelled the mission. The formations seen on the air surveillance radars were false and a result of Hungarian or Soviet electronic warfare that mislead the Yugoslavs – probably in an attempt to find out the reaction time of the RV i PVO. Nevertheless, this sudden test of air capability was remarked upon later as good proof of the morale among the fighter pilots and the combat readiness of the entire air force.[43]

MiG-21PFM No. 22733 seen in QRA attached to an APU vehicle. (Stanko Janjić)

A MiG-21F-13, armed with R-3S AAMs and attached to an APU vehicle (APA-4), seen at its QRA station outside Objekt Klek in 1968. (Suad Hamzić)

A MiG-21 of 117th Regiment in Combat readiness No. 1 during the "Czechoslovak Crisis" of August-October 1968. (Author's collection)

During the "Czechoslovak Crisis", No. 352 Reconnaissance Squadron acted as a fighter unit, equipped with MiG-21Fs. This photograph shows a "flying day" during the crisis. (Novica Šušić)

landing." 204th Regiment was especially tasked and maintained the QRA fighter pair in combat readiness No. 1, a detachment in each squadron in combat readiness No. 2, and one squadron in the regiment and regimental HQ in combat readiness No. 3. At night, 50 percent of the personnel that were in combat readiness during the day remained at their posts. It is almost a paradox that Soviet "Guarantees" technicians that arrived to maintain PFMs remained in Belgrade, despite the crisis.[44]

The "Czechoslovak Crisis" was a serious challenge. In fact, 204th Regiment was the only Yugoslav fighter asset capable of facing the eventual threat of the Soviets or Warsaw Pact. The lack of available interceptors reached such proportions that the fleet of F-84G Thunderjets from fighter-bomber squadrons was assigned the duties of flying combat air patrols (CAPs) and standing QRA at their home bases. The HQ of 15th Division PVO noted that 117th Regiment reached full operational capability by the end of the year probably with only one squadron, No. 125. It is most likely that No. 124 Squadron was still undergoing the training to reach its IOC. No. 352 Squadron was capable of visual reconnaissance in VFR conditions while waiting for the new MiG-21Rs to arrive from the Soviet Union.

117th Regiment was still acclimatising to Bihać Air Base. At the moment of alert, two squadrons of this regiment (Nos. 124 and 352) were in Batajnica in the late stage of conversion to MiG-21F. Not reaching the IOC they were ordered to move immediately to their new home-base at Bihać: No. 352 Squadron took off with its MiG-21F-13s on 26 August. The pilots of No. 124 Squadron were boarded on C-47s and ferried to Bihać. Later, on 12 September, pilots of 204th Regiment ferried 12 MiG-21F-13s and a few MiG-21U-400s to the base. The regiment thus received a complement of 26 single seat and 8 two-seat aircraft, and quickly established a QRA with 'four plus four aircraft' kept at combat readiness No. 1 and No. 2, while all the other aircraft were held at the combat readiness No. 3. This unit lost two pilots in accidents during the crisis. One during the ferry mission from Batajnica, flying a two-seater; the other was lost while on QRA duty and scrambling with a MiG-21F-13.

Exercise Avala finished on 8 October. However, both MiG-21 equipped regiments remained in the higher combat readiness for several weeks after the conclusion of the crisis, providing the "measures… for rejecting the surprise attack from the airspace, especially airborne

Reconnaissance MiGs

The planned delivery of MiG-21Rs began soon after the Czechoslovak Crisis was over. No. 352 Squadron, which was tasked to receive this version, was at Bihać and the first two examples arrived on 28 October 1968, just days after the end of the crisis, and were accepted by No. 126 Fighter Aviation Squadron of 204th Regiment in Batajnica.[45] Nevertheless, the Czechoslovak Crisis hampered political relations between Yugoslavia and Soviet Union. It produced the notion that the Soviet experts (*garantchiki* or guarantees) that arrived with new MiGs, should not be allowed to enter Bihać AB, and even less so Objekt Klek. The further MiG-21Rs (marked initially as L-14R later as L-14i where "i" – stands for *izvidjač* or reconnaissance) continued to arrive from the Soviet Union in July 1969 (six) and April 1970 (four). All of them remained for the one-year guarantee period at Batajnica and then successively joined No. 352 Squadron, until the unit was complete in May 1971.[46] Meanwhile, this squadron operated as a fighter unit with F/PFM airframes, maintaining the QRA pair with Nos. 124 and 125

Seen at Batajnica during the "one year guarantee period", is MiG-21R No. 26107, delivered in July 1969. Its "R" ELINT pod is examined by the cadets of the Air Technical Academy. (Author's collection)

Squadrons. After the delivery of the R version, the squadron converted its 1st Detachment for photoreconnaissance and electronic intelligence gathering missions (ELINT). That said, the squadron retained a dual, fighter-reconnaissance role until 1973, when – although still based at Bihać – it was administratively and operationally separated from 117th Regiment and placed under the direct control of II Intelligence Department of the HQ RV i PVO. Thus, the unit became a strategic asset.

The MiG-21R offered a very wide range of capabilities in gathering aerial intelligence and ELINT. It could be equipped with two major types of underbelly containers: the so-called D-pods and R-pods. The former was capable to detecting and recording the work of enemy radars, and – at least roughly – locating their positions. The latter was equipped with K-24 and similar 'K-series' cameras (acquired by the Yugoslavs during the MDAP-era) for classic aerial photography. The first ELINT-operations by MiG-21Rs of No. 352 Squadron were undertaken on 15 and 16 October 1969 by Captain Stipić along the borders with Italy and Greece. The Yugoslavs were especially proud of getting their MiG-21Rs into operation because the Soviets had rejected the RV i PVO's request to train its pilots and ground crews for such missions. They even refused to deliver the films used by the cameras and recorders, forcing the air force to improvise.[47] Nevertheless, by early 1970, the entire No. 352 Squadron reached IOC status, followed by full operational capability a year later.[48]

2
MIGS OF THE 1970s

Another MiG-21 version was obtained for the RV i PVO in the 1970s. The dynamics of the purchases of the new versions of a fighter are hardly imaginable today in the air forces of the former Yugoslav republics. This model was the MiG-21M (L-15) and a batch of these arrived at Batajnica Air Base and entered the inventory of 204th Fighter Regiment. The M version featured the improved R-11F2SK-300 engine, mounted a GSh-23L gun in the fuselage and had four under-wing pylons for air-to-air or air-to-ground ordnance. Twenty-one of the new aircraft arrived in September 1970, while four more reached Batajnica in January 1972, completing a batch of 25 MiG-21M airframes.[1] The MiGs with serial numbers between 22801 and 22821 had R-11F2S-300 engines, while four later aircraft, 22822-22825, had R-11F2SK-300 engines. In Yugoslav technical documentation, both versions were marked as MM-15.

No. 22801 was the first among the 25 MiG-21M (L-15) delivered to the RV i PVO in 1970-1972. Here it is seen landing with an unusual load of three 490-litre fuel tanks. Later its R-11F2SK-300 engine (Yugoslav MM-15) was replaced with the more powerful R-13-300 (MM-16) and was thus brought to the MF standard. (MC Odbrana)

Six MiG-21MFs were delivered to the RV i PVO in 1975. Here, one is seen at Batajnica with 204th Fighter Regiment. In 1984-85 four of the surviving MFs were converted to carry LORAP photo-reconnaissance pods and were assigned to No. 352 Reconnaissance Squadron. (MC Odbrana)

The new M version was used for the conversion of two squadrons of 204th Fighter Regiment (Nos. 126 and 127), while the third (No. 128) remained with two-seaters and a few PFMs tasked to continue as the conversion unit for all MiG pilots in the RV i PVO. The first unit which would arrive for conversion in 128th Squadron would be the pilots from the 83rd Regiment in Skoplje.[2]

The M version was followed also by the delivery of the advanced two-seater US and early UM versions marked in the RV i PVO as NL-14 and NL-14M, no matter their version differences. Seven of these trainers reached Yugoslavia in two batches that arrived in October 1970 and August 1971. Much later, in January 1975 the L-15 fleet would be completed with the delivery of six MiG-21MFs which would join 204th Regiment.[3] Those featured a further development of the Tumansky engine, the R-13-300. Later, after overhauls, a number of the Ms were brought to MF standard by replacing the R-11F2SK-300 (MM-15) with R-13-300 (MM-16) engines. It was the same practice with early UMs which also received R-13-300 engines after overhaul.

After conversion to the MiG-21M, 204th Regiment was equipped with the most modern type in the RV i PVO inventory. Its two fighter squadrons (Nos. 126 and 127) would train for all types of fighter interception missions and included live gunnery practice over the south Adriatic. The improved capabilities of this version would enable expansion of fighter-bomber/strike training. Now 204th Regiment would be able to carry out strike missions in four-ship formations, which was a significant improvement compared to earlier versions.

Delivery of the MiG-21M enabled further shifting of the airframes and further conversions. The PFM/L-14s were mostly withdrawn from Batajnica's squadrons and now were passed on for the re-equipment of the two fighter squadrons of 117th Regiment in Bihać (Nos. 124 and 125). After two and half years of usage in Bihać, their Fs were passed on to 83rd Fighter Regiment in Skopski Petrovac Air Base.

In the meantime, 117th Regiment had been struggling with severe conditions in Bihać Air Base. Flying of the MiG-21 was hampered by difficulties at the air base itself. Harsh meteo-conditions and lack of the proper landing equipment forced the MiG-21s to be transferred to Zadar Air Base on the Adriatic coast during the winter. This practice continued up to the mid-1970s. The other problem was the accommodation of the technical maintenance at the underground objekt. The result of all of these problems was that the pilots reached between 45 and 65 annual flying hours instead of 80. The unit carried out its first live gunnery practice in 1969 over the south Adriatic with very good results. In the 1970s 117th was converted to the PFM (L-14) version. Again, the conversion was carried out in Batajnica with No. 128 Squadron of 204th Regiment. Now it was much easier than in 1967-68. The differences between F and PFM versions were slight in comparison to the earlier differences between the Sabres and the MiGs. Now both fighter squadrons (Nos. 124 and 125) achieved the same level of operational capability, which had not been the case with the F previously. Unfortunately, the whole period was marked with several crashes that caused the loss of half a dozen pilots.[4]

The problem that was outlined in the late 1960s and by the beginning of the 1970s was the low rate of the MiG-21 usage. Generally, due to the lack of pilots, or the longer pre-flight procedures, the MiGs were used much less than was expected. The economical side of the usage of MiG-21s in 204th Regiment in the 1970s showed that the fleet of MiG-21F-13s carried out around 50 flying hours annually and a total of 250 hours in the whole duration of the usage within the regiment. Soviet estimations for the same period were double and reached between 450 and 500 flying hours. For the two-seater it was an average 263 hours in total and with annual usage reaching up to 87 hours. The Soviet recommendation was 150 annual flying hours. In their first two years the PFMs usage reached an average 88 flying hours instead of the expected 125 hours. The Yugoslavs estimated that the full ratio of MiG-21 usage was possible if the flying day could be organised in two shifts throughout the day, reaching 6-8 sorties per day. It was necessary on average there should be two pilots per MiG

Alongside the M version, Yugoslavia obtained four advanced two-seaters marked as MiG-21US, or NL-14 in the RV i PVO. Here No. 22951 is seen with 204th Fighter Regiment at Batajnica. (MC Odbrana)

Three of the early MiG-21UM version were obtained in August 1971 and marked as NL-14M in the RV i PVO. A decade later they were standardised with a batch of the later produced UM version (NL-16) by switching to R-13F-300 engines. (Author's collection)

and an increase in the number of ground crew and necessary spare parts to enable "economic" usage of the MiG fleet.[5]

Large Exercises

The aggression of the Warsaw Pact armies against Czechoslovakia in the late summer of 1968 and fears within Yugoslavia's ruling circles for the country's independence brought about a rethinking of Yugoslav defence strategies. Now the Soviets and Warsaw Pact were marked again as the possible aggressors. Thus, from late 1968 the defence plans were revised in accordance with such a concept.

The outcome of the development of such concepts was several large exercises that were held in the first half of the 1970s. The basic scenario was taken from the Czechoslovakian "events": sudden aggression, usage of large strike packages and massive armoured columns, inevitable airborne landings in strategic positions or in the rear of the Yugoslav Army, and wide resistance inspired by the Partisan struggle in the Second World War. The fighter component of the air defence, i.e. the MiG-21 fleet, had an important role being the first line of the air defence and as an addition to the strike fleet that would deliver the punch to the aggressors.

Probably the largest manoeuvre ever held in Tito's Yugoslavia occurred between 2

Two of 204th Regiment's PFMs armed with UB-16-57 launchers in May 1969. (Suad Hamzić)

and 9 October 1971. This was "Freedom 71" (Sloboda 71), a manoeuvre held in the Lika – Kordun area in Croatia. Two MiG-21 regiments were engaged well before the manoeuvre in several tactical exercises preparing for this important event. Following the exercise scenario, No. 126 Squadron of 204th Regiment moved from Batajnica to Bihać, while No. 127 Squadron moved to Tuzla. The complete 117th Regiment moved from Bihać to Pleso. Each of the three squadrons was standardised with 10 MiG-21PFMs (or Rs in the case of No. 352 Squadron), and a pair of two-seaters. The manoeuvre was personally inspected by Marshal Tito and the state leadership. The peak of the engagement for 117th Regiment was on 5 and 6 October when the MiGs provided air cover for an assault across the Kupa River and a helicopter landing, and when a formation of 12 MiGs flew an air strike on targets at the Slunj firing range.

Two PFMs (Nos. 22735 and 22733) being prepared for a night sortie during an exercise. (Author's collection)

As the most capable squadron at the time, No. 126 Squadron was tasked to carry out strikes against targets at the large Slunj army range in Lika while Marshal Tito observed the effects from the stage alongside. There were 12 MiG-21Ms, eight of them armed with unguided rockets and four with standard FAB-250M54 bombs. Tito was impressed and all of the engaged MiG squadrons received some kind of prize for their impressive fighter-bomber performance.

Prior to the manoeuvre in each of the units, massive training was organised in fighter-bomber and interception missions. 117th Regiment was converted to use the GP-9 gun-pod on all available PFMs. The full conversion of No. 125 Squadron was completed in pre-manoeuvre conditions and suffered from a lack of personnel, the closure of runaway No. 2, and occasional malfunctions of GCA and other landing equipment. In addition, they still needed to undertake the regular tasks of maintaining QRA at Pula Air Base when Tito was in his residence in the Brioni islands.

A total of 65 MiG-21s of all available versions were used during the manoeuvres, including five squadrons: the only unit not involved was No. 128 Squadron which served as an operational conversion unit

Two scenes from the Sloboda 71 manoeuvres showing MiG-21PFMs at Pleso-Zagreb Air Base. The white lightning flash painted on these aircraft was the symbol of the aggressors during this exercise. (MJV)

MiG-21R No. 26101 seen during pre-flight preparation, during Sloboda 71, at Pleso-Zagreb Air Base. (R.Djilas family via A. Smiljanić)

No. 22808 taxies out onto a runway at Batajnica armed with a UB-16-57 launcher. On 22 July 1976 this aircraft would be lost in fatal crash, killing Bojan Savnik, the youngest RV i PVO general. (Author's collection)

(OCU) at Batajnica and stood the QRA. The MiG fleet maintained almost 100 percent airworthiness, while flying six to seven sorties on average per day. The time norms for arming and launching MiGs were checked during this exercise. A single PFM could be prepared for a mission with two R-3S missiles within six minutes on average. On the other hand, a MiG-21M armed with strike ordnance (UB-16-57 launchers or four FAB-50 or FAB-250 bombs) could be prepared for a mission within 22 and 40 minutes, depending on the ordnance in question.[6]

Podgora 72 and the Crisis with Italy in 1974

In autumn 1972 a large joint exercise was organised in the central Adriatic under the designation Podgora 72. The MiG-21s from 117th Fighter Regiment were engaged in this exercise. Podgora was the first exercise where the MiGs operated in large number over the Adriatic Sea and alongside the coast and islands – which was an entirely new experience for most of the RV i PVO. In this exercise 117th Regiment was not used as a complete entity and it passed No. 124 Squadron to the "Red" side under 82nd Aviation Brigade, and No. 352 Squadron to the "Blue" side under 97th Brigade and deployed it to Mostar. This was the first larger exercise where the reconnaissance MiGs were used for their primary purpose. During Podgora 72 they carried out aerial photoreconnaissance missions over the Adriatic and several mission at the edge of Italian territorial waters and deeper in the Ionian Sea, taking photo imagery of Italian and NATO facilities and ships.

At the end of the exercise, Marshal Tito observed the parade of the Yugoslav Navy in the middle of the Adriatic and the MiGs overflew the missile boats in low-level flight. Upon their return to Bihać, the MiGs were literary salty from the sea water from the low-level pass.

Unresolved border problems with Italy and the presence of the US 6th Fleet in the north Adriatic in spring 1974 was the trigger for another international crisis that hampered relations between Yugoslavia and NATO. The Yugoslav RV i PVO deployed four of its MiG-21 squadrons during the crisis which was run under the codename Plan Koper-Gorica 74.

Moments prior to a planned night flying training flight on 20 March, No. 126 Squadron was alerted to move immediately from Batajnica to Pula Air Base. Pilots boarded their MiGs and made a nocturnal overflight, while the technical company assembled what was needed and proceed to Pula on the ground. On the same morning No. 127 Squadron was alerted and received the order to move to Pleso-Zagreb. No. 126 remained in Pula for two months, while No. 127 in Pleso stayed for a single month. No. 128 Squadron remained with its two-seaters and PFMs in Batajnica carrying out their regular missions, while its pilots occasionally were sent to the two deployed squadrons as replacements.[7]

117th Regiment deployed a detachment on rotational basis to Ljubljana airport. It was a mostly civil airport, and the fighters from Bihać were in tent conditions throughout their deployment. They maintained crews in combat readiness No. 2. The airport fence was constantly crowded by the local Slovenes from neighbouring villages and they had opportunity for the first time to observe the taking-off and landing of the powerful MiGs as Ranko Kovačević, one of the pilots, recalls.[8]

Yugoslav MiG pilots conducted CAP missions in the airspace that borders Italy and over the Adriatic Sea. They also carried out different

Seen here at Mostar Air Base are PFMs of 117th Regiment during the exercise Podgora 72 that was held mostly over the Adriatic Sea. (HDA)

"demonstrative missions" and co-operated with the Yugoslav Navy in the north Adriatic, including some training strike missions on targets at sea. It was intended that the Italians would have no doubt that the Yugoslavs would take a firm stand if any provocations occurred from their side. Most of the missions were coordinated by 15th Air Defence Division and V Air Corps.[9]

Photoreconnaissance, ELINT and SIGINT Missions

Under orders issued in 1973, No. 352 Reconnaissance Squadron left the structure of 117th Fighter Regiment and became an independent unit directly under the RV i PVO HQ and its missions would be coordinated further by its Second (Intelligence) Department. The squadron remained in its premises in Bihać but became a strategic asset of the Yugoslav Air Force. The squadron would maintain an "auxiliary" fighter proficiency, while most of the missions would now be strategic aerial photoreconnaissance and ELINT/SIGINT missions.

The squadron operated a fleet of 12 MiG-21Rs fitted with different underbelly containers: "D" for aerial-photoreconnaissance with six AFA A-39 cameras and a single AShChaFA-5M camera and the R container (marked in RV i PVO as "E" container) for ELINT/SIGINT coupled with one A-39 camera. Container "N" for night aerial photoreconnaissance missions, which had two vertical UA-47 cameras and illumination flares, was delivered much later in 1978. In the usual mission configuration each of the MiGs also carried two under-wing fuel tanks. As Suad Hamzić recalls, such equipment downgraded the aircrafts' capabilities. In full reconnaissance configuration, the maximum altitude which "R" could reach was 14,000 metres and maximum speed at 11,000 metres was Mach 1.3.[10]

Usage of the MiG-21R improved the Yugoslav Army's overall capability to monitor the radar activities of the many neighbouring countries, which prior to the delivery of this aircraft, was carried out by ground-based surveillance stations with variable results.

The "radio-technical reconnaissance" missions were usually carried out on an annual basis for each of the neighbouring countries. The squadron would deploy the one or two MiG-21Rs with ground crew teams to Batajnica, Priština or Titograd Air Base. The mission was carried out by flying near the borders of the neighbouring countries. As most pilots recall the emphasis was given to NATO members Italy and Greece, and then to Warsaw Pact members Hungary, Romania and Bulgaria. Albania and Austria remained out of particular interest for Yugoslav ELINT/SIGINT.

The other type of mission did not involve deployment to the other air bases. The MiG would take off from Bihać, continue along the border of Hungary, then land at Batajnica for refuelling and continue near the borders of Romania and Bulgaria, level with the city of Veles in Macedonia, before making a turn and heading directly back to Bihać.

The missions "against" Italy followed a north-south course between Pula and Titograd Air Base. They first followed a straight course alongside the Adriatic shore, then continued with diving from 10,000 metres heading towards the particular object (radar station) which was the point of interest. Alternatively, from flying a straight course, when level with the object of interest a pilot would make a 360 degree

Because of the harsh weather and lack of landing equipment, Bihać's MiGs were frequent guests in Zadar-Zemunik Air Base in the early 1970s. Here the pilots of No. 352 Squadron prepare for a mission at the flight line. In the background, MiG-21PFM No. 22703 and several Rs can be seen. (Stevo Janjanin)

A MiG-21R being towed to their temporary apron at Mostar Air Base during the Podgora 72 exercise. (Author's collection)

These PFMs seen in one of the galleries of the underground Objekt Klek at Bihać AB in 1975 belong to 117th Fighter Regiment. (M. Mustapić/MC Odbrana)

estimations, diving all the way to the edge of Italian territorial waters, and then returning back in low-level flight. Usually, the object of such missions was the Gargano peninsula. Radio-silence was mandatory for the whole duration of the mission.[11]

One such "combat sortie" was described by Gojko Kondić, at that time a seasoned 352nd Reconnaissance Squadron pilot. Kondić took-off from Bihać in a MiG-21R heading to the Adriatic shore, then passing the Church of St. Donat in Zadar, and maintaining low level flight at an altitude of 100 metres continued for exactly seven minutes over the sea in a south-west direction towards the coast of Italy. In the seventh minute, he made a sudden u-turn and started climbing to 6,000 metres returning back to Bihać. Kondić provoked the ships of US 6th Fleet, which were in the mid-Adriatic, to turn on their radars which were then registered by the Yugoslav SIGINT and ELINT assets.[12]

The other type of mission where the Italian air defence was monitored was organised in a much more complex fashion. Four L-14i would take off from Pula Air Base and fly at different altitudes between 500 and 2,000 metres over the sea. In the meantime, four other MiG-21s would take off from Bihać and continue directly towards the Italian airspace and make a u-turn back. The reconnaissance MiGs would monitor and film the reaction of Italian air defences to the approaching MiGs. In some of the missions, the reconnaissance MiGs were spotted by Italian fighters, as happened to Suad Hamzić in the Gulf of Trieste, or to Stevo Janjanin in the vicinity of Ancona. Pilot Novica Šušić remembered how he entered Bulgarian and Albanian airspace, continued with filming and returned back to base turn and then continue with the mission to another object and the turn would be repeated. The shortened mission would involve taking off from Bihać, climbing to 10,000 metres and using time and distance without being chased.[13]

Later in the 1980s, some weaknesses of the "R" container were discovered. Yugoslav radars had to be turned off during the sortie

Pilots of No. 125 Squadron, commanded by Major Ranko Krivokapić "Šljivka" (standing nearest to the aircraft), pose near MiG-21PFM No. 22704, at QRA in Pula Air Base on 21 July 1976. (Borislav Pejić)

and deciphering took a long time and a number of skilled technicians. The position of discovered "enemy" radars was only approximate and with uncertainty as to the type.

QRA for Tito's residence in the Brioni Islands

During the 1970s an important assignment for 117th Regiment was to maintain a QRA detachment of four L-14s at Pula Air Base, providing air-cover for Marshal Tito's residence in the Brioni archipelago. The QRA alert was carried out during daylight, starting out half-an-hour before sunrise and half-an-hour after sundown, though, of course, only when Tito was present at Brioni. As Hamzic recalls it was "long... exhausting.... In the night we maintain[ed] the combat readiness No. 3 in which [a] two-ship patrol have to be ready to take off in 30 minutes from receiving the alert."

The QRA in Pula was maintained with one pair of MiGs at readiness No. 1 (pilots in cockpit, airborne in up to 3 minutes) and with another pair at No. 2 (pilots beside the prepared fighters, airborne in up to 6 minutes). It was difficult for the pilots to constantly sit in the cockpit in VKK pilot suits and since testing had proved that MiGs could be airborne in two minutes from alert to retracting the landing gear, it was approved that the pilots at Pula should be at ease, maintaining combat readiness No. 3.[14]

There were often light civil aircraft that violated Yugoslav airspace without warning, sometimes engaged in smuggling flights. Occasionally, the Italian Air Force made flights with their F-104 Starfighters up to the edge of the Yugoslav territorial waters, and then made u-turns to test how quickly the Yugoslav air defence system could react. Upon the orders of 15th Air Defence Division similar missions were organised from the Yugoslav side. A pair of MiGs would take off from Pula and continued over the open sea towards the Italian city of Ancona for up to seven minutes, and then the MiGs would make a turn and continued back to Pula. Five minutes later another pair of MiGs would carry out the same mission, as one of the pilots, Ranko Kovačević, explained. Such Yugoslav responses reduced the Italian intrusions. The practice of maintaining the QRA at Pula Air Base by 117th Regiment continued until 1979, and Marshal Tito died in May 1980. Sometimes, when the conditions for flying in Bihać were difficult due to snow or heavy storms, the regiment deployed flights to Pula and Zemunik, where they continued training and flying.[15]

83rd Regiment and Priština AB

The remaining ex-117th Regiment Sabre Dogs from Pleso-Zagreb were ferried to Skopski Petrovac AB, near Skoplje, in several batches from March 1968. They entered service with No. 123 Fighter Aviation Squadron that had previously operated F-86E Sabres. Conversion on the Sabre Dogs lasted throughout 1968, especially for the usage of the radar. The Czechoslovak Crisis occurred in the middle of the squadron's conversion. The senior leadership from the RV i PVO decided to expand No. 123 Squadron into a regiment, adding some more Sabre Dogs from reserves in Zagreb and a group of 20 young pilots, second lieutenants that had finished academy just prior to the crisis. By the beginning of October 1968, a new 83rd Fighter Regiment was formed. Beside No. 123 Squadron, this regiment established another fighter squadron, No. 121, which soon changed its designation to No. 130.[16]

83rd Fighter Aviation Regiment would be the third Yugoslav regiment to operate MiG-21 fighters.[17] Bihać's pilots ferried their MiG-21Fs to Skoplje in February 1971. Both squadrons of 83rd Regiment carried out their conversion during 1970-1971 at Batajnica and Skoplje, starting with No. 130 and followed by No. 123. On 20 March 1971, this regiment entered the QRA system, responsible for the south-eastern part of Yugoslavia.[18]

83rd Regiment remained in Skopski Petrovac until October of 1972. It briefly moved to Titograd (Podgorica) where it was engaged in air exercises from 15-17 October. After landing at Skoplje on the afternoon of 17 October 1972, commanding officer Lt. Colonel Angel Ončevski received a telegram with orders that on the next morning his Regiment had to move to a new air base near Priština. Despite the exhaustion of the personnel that had just arrived from exercise over Montenegro, Ončevski ordered that the personnel should continue their activities and prepare for the movement to the new air base. On the morning of 18 October 1972, the first MiGs of 83rd Regiment landed on Priština's runaway.[19] The technical segments of the regiment continued to arrive up until February 1973.

The new air base near Priština had an underground objekt, code-named "Rudnik", for the MiGs. For taking-off, two auxiliary runaways that led from either side of the underground objekt could be used. The base was still unfinished and conditions in the new air base were harsh. As General Angel Ončevski, that time commander of 83rd Regiment points out "It was rush, without need!" Priština had never before been an air base. Most of the personnel lived in the air base premises regardless of whether they were officers or NCOs as the promised housing was yet to be constructed. Night and IFR landing equipment were not yet ready and some AA artillery searchlights were borrowed from the ground forces for night flying.

Flying at Priština was very much a demanding task as most of the pilots recall. The air base was at 550 metres above sea level which meant that pilots needed to use higher speeds in taking off and landing. The approach was difficult due to the mountain ranges that surrounded the plains of Kosovo.

A total surprise for the pilots and senior leadership in the RV i PVO HQ was the dust which came out of the chimneys of the Oblić power plant when the wind blew from the northern side. On the "dry" days it was not a huge problem, but on the rainy days, the dust that covered the runway, taxiway and aprons turned all of them into slippery surfaces where aircraft would slide as though on oil. It was not unusual for some of the MiGs to slide from the runaways into the nearby mud and this problem remained a constant during the whole period of the usage of Priština Air Base.[20] Commonly – though never officially – known as Slatina, after a village north of the runway, this was a name was not easily forgotten.

Seen at Skopski Petrovac Air Base during pre-flight checks is No. 22533 belonging to 83rd Fighter Regiment. (83.lap)

MiG-21F No. 22523 of 83rd Aviation Brigade returns to the apron during a tactical exercise. (83.lap)

83rd Aviation Brigade – a Yugoslav Air Defence Experiment

The movement of 83rd Regiment to the new air base near Priština led to organisational change. Since the paternal 13th Air Defence Division had not been established in full with all of its components, it was officially disbanded. 83rd Regiment was then reorganised as 83rd Air Defence Brigade by orders issued on 6 October 1972, adding to the existing two fighter squadrons 31st Air Surveillance Battalion, a Combat Command Group and a Staff Detachment equipped with helicopters. One of the radar companies of 31st Battalion was equipped with the British-made Marconi S-600 radar and positioned on the nearby Goleš Mountain, while the battalion's operational centre was located in the underground Objekt Rudnik, thus forming a theoretic unity of fighter aviation and radar network and control.[21]

The creation of 83rd Brigade was something between a necessity, an improvisation and an experiment in the field of organising Yugoslav air defence. It was intended that this kind of brigade would be an answer to the problems of linking fighter aviation and air surveillance systems into one automated system. The intention was to link as closely as possible the information that radars detected, the QRA and the guidance of the fighter aviation to the target.[22]

Soon, practice showed that the Brigade HQ maintained focus on the fighter squadrons since it was actually the previous HQ and staff of the fighter regiment. The air surveillance battalion had gradually been neglected since it was added to the aviation unit and separated from other air surveillance units in all other aspects from maintenance to human resources. It was obvious that it was difficult to maintain such a joint tactical air defence unit.[23] By late 1976 it was decided to disband the brigade as soon as it would be possible. By the beginning of 1977, the HQ of 11th Air Defence Division ordered that 83rd Brigade should operate as the earlier two-squadron fighter regiment, while 31st Air Surveillance Battalion should be detached to 1st Air Surveillance

The loading of a UB-16-57 launcher with S-5 57mm unguided rockets under the wing of MiG-21F No. 22507 of 83rd Fighter Regiment/Aviation Brigade, at Priština Air Base in the mid-1970s. (Author's collection)

Colonel Branko Gajević "The Frenchmen", commander of 83rd Aviation Brigade poses near the U-model two-seater No. 22902. (Gajević family)

Regiment. Finally, by the orders of the Federal Ministry of Defence and RV i PVO HQ, dated 13 December 1977, 83rd Brigade was returned to the earlier organisational status of a fighter regiment with two fighter squadrons.[24] We can conclude that such a joint tactical air defence unit was impossible to function and maintain in the Yugoslav practice. From 1 April 1978, 83rd Regiment was subordinated to the 11th Air Defence Division, as was 204th Regiment.

Table 3: MiG-21s of the RV i PVO, 1 February 1975

Fighter Regiment	Squadron	Available aircraft and versions
204	HQ Det	1 MiG-21M, 2 MiG-21PFM, 1 MiG-21US
	126.	14 MiG-21M, 1 MiG-21US
	127.	13 MiG-21M, 1 MiG-21US
	128.	4 MiG-21PFM, 4 MiG-21U, 2 MiG-21US
117	HQ Det	3 MiG-21PFM, 1 MiG-21U
	124.	10 MiG-21PFM, 2 MiG-21U
	125.	8 MiG-21PFM, 2 MiG-21U
	352.	11 MiG-21R, 1 MiG-21US
83	HQ Det	2 MiG-21F-13
	123.	13 MiG-21F-13, 2 MiG-21U
	130.	12 MiG-21F-13, 3 MiG-21U
Zmaj Air Depot		1 MiG-21F-13, 5 MiG-21PFM, 1 MiG-21M, 1 MiG-21R 1 MiG-21U
Total		29 MiG-21M/MF, 32 MiG-21PFM, 12 MiG-21R, 28 MiG-21F-13, 6 MiG-21US, 15 MiG-21U

Overhauls

Overhauling of the MiG-21s was organised at Zmaj Air Depot at Velika Gorica, on the eastern side of Pleso-Zagreb Air Base. This institution started to overhaul MiG-21Fs and Us in late 1967, marked as the "middle life repair", since the technicians who were trained in the Soviet Union had yet not returned to Yugoslavia.[25]

The process of "conquering" the overhauling lasted for the next few years. Another problem was that the Soviets closed the producing line for MiG-21Fs in 1968 and for PFMs in 1970. The Yugoslavs estimated that spare parts would be a problem in years to come and soon learned that, unlike the US MDAP-supplied aircraft, the major parts of the MiG-21s were not interchangeable between the airframes. Each part would fit only to its own aircraft and it was sometimes impossible to accommodate some of the parts when changing between the aircraft.[26]

By the beginning of the 1970s Zmaj started to carry out overhauls of the PFMs. In 1975 they started to overhaul the M version, in 1978 a two-seater US version, and by the beginning of the 1980s bis and UM two-seaters. During the 1970s it mostly took up to two years for overhaul, while in the mid-1980s it reduced to between 11 and 14 months.[27]

By the beginning of the 1980s, Zmaj also started to overhaul foreign MiG-21s, mostly from Third World air forces including Sudan and Egypt. By offering the cheapest price of US$8,300 the Yugoslavs received the "Egyptian job".[28] However, Zmaj lost the contract with Egypt when the East Germans offered even lower prices. In 1989 a contract with Iraq was agreed to overhaul 30 MiG-21bis (in the Yugoslav official documents the version is marked as SMT, or as MiG-21IQ even L-17IQ) and 10 MiG-23MLs. These were ferried aboard Il-76 transporters and passed on to the overhauling line alongside the Yugoslav bis/L-17s in 1990-1991.[29]

The Orao Air Depot at Rajlovac near Sarajevo was responsible for overhauling the jet engines for the RV i PVO aircraft. In 1966, the Yugoslavs sent a small quantity of Tumansky engines to the USSR for overhauling and later a group of technicians to train to establish an overhauling line in Yugoslavia. The engines remained in the Soviet Union and the overhauling was further prolonged. Overhauling of the Yugoslav R-11F-300 engines started by the beginning of 1970. In 1968 it was estimated that the overhaul of the engines would be critical. The Soviets refused to deliver the test station for engines and so the Yugoslavs had to develop their own. It was estimated that

One of the first MiG-21s that entered the Zmaj Air Depot in 1967 was the U version serialed 22902, seen here between two F-86D Sabre Dogs, making a vivid illustration of the East-West symbiosis within the Yugoslav fighter aviation of the 1960s (HDA)

The overhaul line for the MiG-21s in the Zmaj Air Depot in the mid-1980s. (Author's collection)

A close look at the overhaul line at Zmaj Air Depot reveals works on MiG-21M No. 22815 and MiG-21bis No. 17129 during the mid-1980s. (Author's collection)

overhauling of the engines would start in 1971 in the Orao Air Depot at Rajlovac.[30] This finally enabled the overhauling of further versions of the Tumansky engines in Yugoslavia.

The engine was marked in the RV i PVO as the MM-12 denoting *Mlazni motor-12*, or Jet Engine 12. Later versions kept the "MM" prefix and the number matched the L-version designation (R-11-F2S-300 or MM-14, R-11-F2SK-300 or MM-15, R-13-300 or MM-16, and R-25-300 or MM-17).

In the earliest phase of usage a percentage of typical MiG incidents were caused by tyre bursts, and problems in the engine which caused flame bursts. An important discovery in 1970 was that Dunlop tyres produced in India had a much higher percentage of incidents, caused by the higher percentage of carbon and their usage was soon forbidden.[31]

The only MiG-21 painted in a camouflage scheme was PFM No. 22719. The camouflage consisted of a standard Yugoslav scheme (inspired by the RAF), of dark green and mid-grey on upper surfaces and sides, and PRU blue on undersurfaces. It was applied by the Zmaj Works in 1979 when the aircraft served with 117th Regiment HQ Detachment and No. 125 Squadron. Photographs show the aircraft sometimes wore white and sometimes black tactical numbers (or 'squadron numbers' in RV i PVO jargon). However, the camouflage was not approved and the aircraft was soon returned to its standard metal finish. (Ivan Nikolovski)

3

BACKBONE OF THE AIR FORCE

The senior leadership of the RV i PVO was planning the introduction of the modern version of the MiG-21 from the early 1970s. Originally, this was provisionally designated as the 'L-X', or even 'MiG-23', and was expected to be capable of carrying out interception tasks – and even fly dogfight-style air combats at high altitudes.[1] As so often, the reality was entirely different. Through the early 1970s, it became plainly obvious that the ground attack capabilities of the air force were in rapid decline. The Soko Jastreb light strikers had only limited capabilities, while the Soko Orao was seriously delayed. Lacking alternatives, the Yugoslavs were de facto forced into acquiring and then deploying the MiG-21bis as multi-role fighter-bombers, rather than as interceptors.

The first MiG-21bis arrived in Yugoslavia in August 1977 and received the local designation L-17. This variant differed from all the previous ones and instead of offering an improved high-altitude performance, as expected, it foremost offered an improved performance at low altitudes. Powered by the new R-25-300 (MM-17) engine with two afterburner stages, it was equipped with the RP-22SMA Saphir-21 radar that supported the deployment of the new R-3R semi-active radar homing air-to-air missiles, the SAU-23 autopilot and the RSBN-5S radio-navigation system, larger wheels for the main landing gear, and other improvements. As such, it offered significant potential as a fighter bomber.[2] Moreover, deliveries of the MiG-21bis were followed by those of much improved MiG-21UM two-seat conversion trainers, powered by the R-13-300 (MM-16) engine.

An excellent photograph of MiG-21bis No. 17169 of 204th Fighter Regiment seen during QRA trials where it was launched from the HAS behind Batajnica's runways. It features newly delivered R-13M and R-3R AAMs received alongside the bis version. (Milan Micevski)

Winter "flying day" at Batajnica, in the mid-1980s reveals the MiG-21bis-K (L-17K) delivered in two last batches in the spring 1982 and spring of 1983. (Author's collection)

A MiG-21bis (L-17) seen in underground Objekt Klek, near Bihać, at the head of a line of No. 124 Squadron airframes in Gallery "C". This aircraft was lost in a non-fatal crash on 9 September 1987. (N. Maravić family)

continued. A group of Soviet technicians remained in the regiment until 1983 to monitor the new fighters in the period of guarantee and to assist ground crew of 204th Regiment.[4]

Headed by 204th's CO Colonel Mirčetić, a group of five pilots went in June 1978 to Astrahan in the USSR for live practice using the R-3R and R-13M missiles on the MiG-21bis. The Yugoslavs asked that the conditions for launching should be as realistic as possible with targets at maximum distance, which manoeuvred, and at the lowest level of the radar. Only one Yugoslav managed to hit the target.[5]

The introduction of the MiG-21bis improved the training in interceptions, which now included daytime in all kind of weather conditions as well as in the night in VFR conditions up to 10,000 metres. Interceptions were classical, semi-automatic, and in the free hunt, as well as from the QRA both on the ground and in the air. Moreover, No. 126 Squadron started to carry out training in strike missions. In the following year, the training expanded the capabilities of the regiment on the bis. Most of the capabilities were demonstrated in several tactical exercises organised in 1978-1980. Among many of the novelties were exercises with operating/taking-off from grass airstrips, and live gunnery practice carried at Kamenjak range, south of Pula in Istria, at the edge of the tactical radius, with direct overflight from Batajnica and back. In 1980, for the first time, there was live practice in Yugoslavia with two MiGs firing single R-3S against a launched RM-3 target, and flying in 15-ship formations.[6]

No. 126 Squadron was the first unit to receive the bis and in 1977 four of its pilots and 22 technicians went to the Higher Military Aviation School (Краснодарское высшее военное авиационное училище летчиков), at Krasonodar in the Soviet Union, for a 45-day conversion course. After the return of the group, a conversion programme in Batajnica was organised.[3] 204th Regiment's HQ and staff and the remaining pilots and technicians of No. 126 Squadron went on this course. The conversion lasted until 1978. It was followed by No. 127 Squadron which completed in the first half of 1979. The conversion in 1979 was slowed because of bad weather and accidents with fatalities. It was supported by some dozen or so Soviet technicians that were attached to the unit. Meanwhile, the regiment maintained QRA and regular activities with MiG-21M/MFs, which started to be phased out to 83rd Regiment as the deliveries of the bis-version

It is important to note that there were differences in the batches of bis (L-17) which were delivered to Yugoslavia in this period. The bis fighters which arrived in 1977-1979 were fitted with the RSBN radio-navigation system and lacked the LAZUR-M guidance system.[7] These were much more fighter-bombers than interceptors and are easily recognisable from the digits on the fuselage ranging from 101 onwards.

In the 1980, another subversion of the bis, now with the LAZUR-M guidance system, started to enter the inventory. This batch was marked in Yugoslavia as the L-17K, and had digits starting from 201 or 401. In 204th Regiment, No. 126 Squadron operated the RSBN-equipped bis version, while No. 127 Squadron the LAZUR-M version.[8] Such practice also later spread to 117th Regiment. Nevertheless, by the late 1980s the Yugoslav L-17 and L-17K versions were merged in the squadrons no matter their initial differences. Contrary to some other practices and interpretations in Yugoslavia these types were simply referred to as L-17 and the name L-17K only appears in early stage documents.

In June 1978, the bis/L-17 started to join 117th Fighter Regiment in Bihać. The first squadron to convert to this type was No. 124 and the Regimental HQ Detachment. The conversion was organised in Batajnica, first for the pilots from Batajnica and No. 125 Squadron carried out conversion in 1979-1980 to this type. Nearly all of the PFM/L-14s from Bihać were passed to 128th Conversion Centre in Batajnica for training of the younger pilots.[9]

Table 4: MiG-21s of the RV i PVO, 31 August 1979

Fighter Regiment	Squadron	Available aircraft and versions
204	HQ Det 126. 127.	3 MiG-21bis, 1 MiG-21UM 16 MiG-21bis, 2 MiG-21UM 14 MiG-21bis, 6 MiG-21M/MF, 2 MiG-21UM
128 Conv. Centre		9 MiG-21PFM, 5 MiG-21U, 8 MiG-21UM
117	HQ Det 124. 125.	4 MiG-21UM, 2 MiG-21U 12 MiG-21bis 16 MiG-21PFM
	352.	11 MiG-21R, 2 MiG-21US
83	HQ Det 123. 130.	2 MiG-21M, 1 MiG-21US 11 MiG-21F-13, 4 MiG-21U 17 MiG-21M, 3 MiG-21US
Air Test Centre		1 MiG-21U
Zmaj Air Depot		2 MiG-21PFM, 1 MiG-21U
Total		45 MiG-21bis, 25 MiG-21M/MF, 27 MiG-21PFM, 11 MiG-21R, 11 MiG-21F-13, 17 MiG-21UM, 6 MiG-21US, 13 MiG-21U

The Preparation of MiG-21bis No. 17128 at the apron of Bihać Air Base in around 1980. (Author's collection)

Fuelling of MiG-21bis No. 17128 during the harsh early winter days of 1980-81 at Bihać Air Base. (Author's collection)

Alert Caused by Marshal Tito's Illness

During the lasting illness of Marshal Tito from autumn 1979 to spring 1980, the Yugoslav armed forces were kept at the higher combat readiness. It was expected that when he passed away, a "possible foreign invader" may suddenly launch an aggression against Yugoslavia. This estimation made no distinction between NATO and the Warsaw Pact. The Air Defence system especially was kept at high readiness, including the MiGs at Batajnica and Bihać. At 117th Regiment in late January 1980, 25% of personnel were kept in constant combat readiness, while others had extended working hours from 06:00 to 18:00. Night missions brought the working hours up to 16 hours per day, though the atmosphere was positive and personnel carried out their tasks with enthusiasm and without complaint. The regiment reached its highest number of flying hours per pilot in this year and maintained a 92.5% percent airworthiness of its MiGs. Some of the experienced pilots, such as Major Mihovilović, CO of No. 125 Squadron, reached 120 and 140 flying hours in this period (1979-1980). The conversion of No. 125 Squadron to the MiG-

21bis, was carried out during this period and it improved the number of sorties conducted.[10]

When Tito died on 4 May, combat readiness reached its peak in the following days until his burial. But nothing happened – nobody invaded Yugoslavia. For the burial ceremony it was ordered that 204th Regiment organise a flypast over the ceremony in Belgrade with two groups of 12 MiG-21s. The task was enormously demanding and important. There should be no mistakes in flying over the capital city. The weather was difficult, clouds were extremely low and the taking-off was in groups of 12 aircraft. The rain made taking-off extremely dangerous, since the first two aircraft in a formation raised a water curtain which remained as each pair in the formation continued to take off. The visibility was absolutely zero. Nevertheless, the Batajnica MiGs made a flyover of Belgrade and saluted their passed Supreme Commander.[11]

A flight-line at Priština around 1980 with MiG-21Ms and a single MF (third in row). After the overhaul, No. 22816 (second in line) had its engine changed to R-13F-300 (MM-16) and was brought up to MF standard. (Author's collection)

Two-seaters of 83rd Fighter Regiment at Priština in 1980. In the centre is No. 22903 belonging to the first batch of the early "U" version. On its right side is an advanced "US" version, No. 22952. (Author's collection)

83rd at Priština is Re-equipped

As was the case in 1970-1971, when the new MiG-21bis were introduced into the inventory this enabled further movement of the airframes elsewhere. Phased out from 204th Regiment's inventory, the M/MFs were ordered to equip 83rd Regiment in Priština. Conversion of the HQ and No. 130 squadron was carried out with ground crew between May and June 1977, while flying training was finished in 1978. No. 130 Squadron was starting to re-equip with M/MFs by the beginning of 1978 when the batch of M/MFs arrived from Batajnica. The regiment "sharply strengthen[ed] its firepower" was the conclusion of the HQ.[12]

A batch of MiG-21F13s was assigned to the Air Technical Academy in Rajlovac to be used for training young cadets, as seen here displayed at Sarajevo-Butmir in May 1987. (Milan Micevski)

Deliveries of M/MFs (L-15) to Priština replaced the ageing F-13 (L-12). Despite being gradually outmatched, the F-13s were still used to full capacity, including live gunnery practice with AAMs over ranges in the mid and south Adriatic and maintained above 80 percent airworthiness. By the order of the Federal Ministry of Defence on 11 April 1981, this type was phased out from the RV i PVO service. In May 1981, a batch of 11 L-12s and NL-12s were flown from Priština to Sombor and Rajlovac were they were re-qualified into training devices for soldiers and cadets of the air technical branch. Now both squadrons of 83rd Regiment operated the M/MF version.

As the deliveries of the MiG-21bis continued, soon 83rd Regiment started with another conversion. No. 123 Squadron was ordered in November 1981 to convert to the bis and the instructors arrived from Batajnica. Besides the HQ and No. 123 Squadron, all of the older pilots from No. 130 Squadron were also converted to the type. But since the bis version deliveries ceased in 1983, this regiment remained equipped with one bis (L-17) squadron and one M (L-15) squadron throughout the 1980s and until 1994.[13]

No. 123 Fighter Squadron of 83rd Regiment was re-equipped with the MiG-21bis in 1981. (83.lap)

Ground crew of 128th Conversion Centre in Batajnica, around 1980, clear the heavy snow to enable their MiGs to fly. Behind the men are four PFMs and seven brand new UM trainers. (Author's collection)

MiGs Join the Military Air Academy

In 1977, the third squadron of 204th Regiment, No. 128 – which was used as the training and conversion unit during the 1970s – was separated from the regiment to form an independent entity. This would become 128th Conversion Centre – for Supersonic Aviation Pilots (Centar za preobuku – pilota supersonične avijacije). The Centre was commanded by one of the most experienced MiG-21 pilots; Lt Col Ante Šutalo.[14] This Centre would be responsible for the conversion of young pilots that had finished Air Academy and were chosen to become MiG pilots, in addition to other pilots who had been on different MiG-21 conversions or just to maintain flying proficiency. By separating this unit, 204th Regiment was devoted solely to combat training with its two fighter squadrons (Nos. 126 and 127).

This unit had the important task of training a group of pilots from the Palestinian Liberation Organization in 1981-82. It was the outcome of the then current Yugoslav foreign policy to support liberation movements in the co-called Third World. The group of Palestinian pilots, led by Major Muhamed Gazzaz, had previously finished basic training at the Libyan Air Force academy and three of them had already flown in Libya on MiG-21s at Kufra Air Base. They were converted on MiG-21PFMs at 128th Centre in Batajnica, but due to the bad weather in the winter of 1981 they continued training at Pula.[15] 128th Centre was supposed to train more groups to fly on the MiG-21bis within 204th Regiment in 1982.[16]

Palestinian pilots, lieutenants Mustafa Jawad and Rabia Ahmed, pose near a MiG-21UM with their Yugoslav instructor Major Savo Jovanović at 128th Conversion Centre at Batajnica in 1981. (Savo Jovanović)

No matter that the separation of 128th Conversion Centre from its former parent regiment relieved the HQ of additional obligations, it was soon discovered that the conversion of large groups of young pilots that arrived from the academy was difficult to conduct in the busy air base of Batajnica. So, it was decided that the Centre should be moved to the Adriatic coast as was the case with other Military Air Academy assets. This decision coincided with further delivery batches of bis-K and two-seater UMs. These deliveries enabled 128th Centre to return to its earlier fighter squadron status, equipped with the regular bis and 10 brand new bis-Ks delivered in May 1983, with conversion on the type carried out until September of the same year.

At Pula Air Base at the southern tip of the Istria Peninsula in 1981 No. 129 Fighter Squadron was formed as part of 185th Fighter-Bomber Aviation Regiment that belonged to the Military Air

From 1981 the main conversion unit for MiG-21s became No. 129 Fighter Squadron of 185th Fighter-Bomber Regiment (Training) in Pula. Here, UM two-seater No. 16155 is seen during landing in 1986. (MC Odbrana)

The other type that was used in No. 129 Squadron was the MiG-21PFM. One of these, No. 22704, is seen here prior to a strike mission armed with UB-16-57 launchers. (Miroslav Špica)

An unusual incident occurred in 1986 when Second Lieutenant Beno Trap hit a hangar at Pula Air Base while taxing towards the runway. It caused the loss of MiG-21PFM No. 22721. (185.lbap)

An impressive view of the ramp of No. 129 Squadron at Pula Air Base, with total of 18 MiG-21PFM/UMs. (Miroslav Špica)

Academy.[17] The new conversion unit was established with a batch of new UM (NL-16) two-seaters and the remaining PFMs (L-14s) which were available in sufficient quantity.

No. 129 Squadron existed in the 1960s equipped with the F-86E Sabre and acted as the conversion hub for the young fighter pilots and cadets in the finishing year of academy. Now the squadron was reformed and entered the system of training the cadets in the Air Academy. In the fourth year of their schooling, the cadets chosen for fighter aviation were passed to this unit for their maiden flights and further conversion to MiGs. The first group of cadets finished this program in 1983 and it continued until 1991-92 when the last group of cadets were converted to MiGs during the war that broke out.[18]

Concepts and Problems of the Yugoslav Fighter Aviation in the First Half of the 1980s

Marshal Tito's death marked an important step in the history of socialist Yugoslavia. He was the cohesive factor for both the state and the Communist Party and was a recognised international player. Tito's departure from the historical and political scene would soon lay bare the problems in the functioning of Yugoslavia. First there were economic problems, which burdened everyday life in the first half of the 1980s. These would be followed by political and national problems that would escalate in 1991 into full-scale hostilities and war among the Yugoslav nations.

In many of the contemporary estimations, the economic crisis in Yugoslavia was regarded as an obstacle for further development of the Yugoslav air defence system by the beginning of the 1980s. But, it was actually an intensive period with planned purchase of the latest MiG-21 versions which led to a rise in the number of MiG-equipped squadrons to a total of nine, with strenuous training in fighter and strike missions and in intensive flying in all weather conditions.

The fleet of MiG-21s in this period counted around 160 combat fighter and reconnaissance aircraft, among them around 85 belonging to the most modern bis version. It was planned that each of the combat fighter squadrons would have 16 aircraft and 26 pilots. The total number of MiG pilots in the units was actually 211 and this meant that there were an average of 1.3 pilots for each MiG; the intention was to improve this number to an average of 1.6:1, which was regarded as the NATO average number of available pilots. It was estimated that besides standard fighter pairs held in QRA, a flight of six to nine fighters in each regiment could be airborne in 15 minutes. A regiment sized group of around 30 fighters could be in the highest combat readiness in one hour, and the complete assets of RV i PVO fighter aviation in three hours from the moment when the alert was given. In 1985, RV i PVO planners estimated that the Yugoslav fighter fleet could perform around "550 sorties per day with [an] effect of between 60 to 70 shot down (enemy) aircrafts!"[19]

Night flying under the searchlights; mid-1982. This particular photo was taken during the making of the film *The Sky is Far* (*Daleko je nebo*). (Aleksandar Smiljanić collection)

The usual QRA load before the new R-60 AAMs were obtained was as seen here: R-13M and R-3R under the pylon of MiG-21bis No. 17169, at Batajnica in May 1987. (Milan Micevski)

In the same period, the Air Defence senior leadership identified the following problems with the fighter aviation: lack of planning number of interceptions guided by the air surveillance units, a low number of planned interceptions which were guided by the air surveillance units, interceptions at lower altitudes, training in free hunt, and an occasional decline in the efficiency of the fighter squadrons when young pilots from the academy joined the units.

In the many years of MiG-21 usage the Yugoslavs had insisted on interception and dogfighting at above the speed of sound. Many of the pilots logged numerous flying hours above Mach 1 no matter the flying difficulties in such circumstances. One of the most experienced Yugoslav test pilots, Franc Hudomal, was surprised to discover that many of the experienced NATO pilots didn't train in supersonic flight or for the conduct of aerial combat above Mach 1, explaining that their F-4s or F-16s were unstable and shaky above Mach 1. Contrary to them, Hudomal and most of the experienced Yugoslav MiG-21 pilots had logged several dozen supersonic interceptions.[20]

Ground crew secure an RM-3 launched target under the wing of a MiG-21bis of 117th Fighter Regiment at Bihać. Used for live practice, the RM-3 was launched prior to the firing of the AAMs. (via B. Pejić)

The fighter units worked hard to match the expected criteria. In 1981, 117th Regiment performed interception at high altitudes, at the operational ceiling, and at supersonic speeds while classically guided by a radar station. The core of the training in 1985 was interception guided by the Air Surveillance Regiment operational centres, with 50% of interceptions carried out at high altitudes. Radar guiding-officers were criticised for mistakes in their guidance to targets at high altitudes and weak contact with pilots in the daily VFR exercises.

Generally, the radar guided interceptions were not carried out in the numbers needed or planned for.[21]

It is obvious that between 1985 and up to 1991, the training in the "auxiliary" (strike) missions prevailed over the air defence training and became much more important in the usage of the MiG-21s in the Yugoslav Air Force.

New Ordnance Expands MiG-21 Strike Capabilities

Since the early 1970s, the Yugoslavs tested different solutions for equipping their MiG-21PFM as strike platforms. They asked the Soviets if they could test the H-66 guided AGM (ASCC/NATO-codename AS-7 Kerry) and two of the pilots and some ground crew were sent in March 1974 to 715th Training Regiment in Lugovaya in the USSR where they each launched a single H-66. One of the pilots, Lt. Colonel Nedin, explained later that the pilot had to maintain visual contact on the rocket in its flight to the target which was regarded as a serious problem in real combat. Since the Soviets requested US$17,760 each, the Yugoslavs abandoned any ideas to equip their MiGs with the Soviet AGMs. The idea of turning the MiG-21PFM into a strike platform continued with deliveries of MiG-21bis which was much more suitable for such missions.

The tactical capabilities of the MiG-21bis enabled its usage as a potent strike platform. Because of the Yugoslav policy of trying to remain between the two major blocs, the RV i PVO obtained several new ordnance types in the United States, Great Britain and France, such as the AGM-65 Maverick, Mk-82 bombs, Hunting BL.755 cluster bombs, and Durandal anti-runaway bombs, and developed several new indigenous types of aerial ordnance, mostly for usage with the Yugoslav-made Orao and Super Galeb light strike combat aircraft. Some of these types were integrated on MiG-21bis.

Most important was the British Hunting BL.755 cluster bomb, the first batch of which was delivered on 29 December 1978.[22] Since 1983-84, gradually all of the MiG pilots were converted to drop this bomb by using basic manoeuvres over the range. In the same time-period, strike training was expanded by using Soviet S-24 unguided rockets and several Yugoslav types of ordnance: L-128-04 launchers with M-74 Munja unguided rockets (first delivered in March 1979),[23] PLAB-350 napalm-bombs, and FAB-250 standard bombs. MiG pilots also trained in the usage of cannon in strafing, unguided bombs, UB-16-57 launchers and even HVAR 5-in unguided rockets on training and tactical ranges. All of the pilots were trained in the use of the standard set of strike ordnance and two-thirds of them for the newly delivered BL.755, S-24, PLAB, and Munja types.

Early version MiG-21UM (NL-14M) and MiG-21M (L-15) of 204th Fighter Regiment, seen at joint tactical exercise Golija 76. Of note are the front part of the fuselage, centreline fuel tank and wingtips pained in light blue, and Munja unguided rocket launchers which were tested during this exercise. (Author's collection)

This aircraft, MiG-21PFM (L-14) serial number 22731, was delivered to Yugoslavia in 1968 and served with 204th Fighter Aviation Regiment until 1978. Subsequently, it was re-assigned to 128th Conversion Centre at Batajnica AB until 1982, when it was re-assigned to 129th Fighter Aviation Squadron (185th Fighter Aviation Regiment) at Pula AB. During the war with Croatia, this unit moved to Tuzla AB in October 1991 where it remained until withdrawn to Serbia in May 1992. Subsequently, this aircraft was assigned to the Priština-based 130th Fighter Aviation Squadron until 1994 and was scrapped two years later according to the Dayton Agreement. It is shown armed with UB-16-57 pods for 57mm S-5K unguided rockets installed on the underwing pylon and a GP-9 gun-pod under the centreline. While its big, black 'tactical number' was removed in October 1991, it retained its 'silver grey' overall livery, consisting of two layers of clear lacquer mixed with aluminium powder. (Artwork by Tom Cooper)

MiG-21R (L-14i) serial number 26107 served its entire carrier – from delivery in July 1970 until mid-1997, when it was scrapped under the Dayton Agreement – with 352nd Reconnaissance Aviation Squadron. In the late 1980s, it was re-engined with a Tumansky R-11F2SK-300 engine (MM-15), like most MiG-21MFs. The aircraft is shown with the D-pod under the centreline: this was a container for photographic cameras. A SAB flash bomb (used to light the target during nocturnal operations) is shown installed on the inboard underwing pylon. Inset in the lower left corner is the R-pod (used for ELINT purposes), and to the right is the standard 490-litre drop tank nearly always carried on outboard underwing pylons. (Artwork by Tom Cooper)

MiG-21bis-K (L-17K) serial number 17407 was assigned to 128th Fighter Aviation Squadron of 204th Fighter Aviation Regiment at Batajnica AB, from its delivery to Yugoslavia in 1983 until 1990. It was then with 125th Squadron of 117th Fighter Aviation Regiment at Bihać AB from 1990 until 1992. The last years of its service were with 124th Squadron of 83rd Aviation Brigade (later Fighter Aviation Regiment) at Ponikve AB. It survived the Kosovo War of 1999 while undergoing overhaul at the Moma Stanojlović Works at Batajnica AB. In 2004, an indigenous reconnaissance pod was tested on this aircraft, but this never entered operational service. Instead, this MiG-21bis-K was one of the last to be used operationally by the Serbian Air Force up until autumn 2015. The aircraft is shown wearing the usual 'air superiority grey' overall and is armed with British-made Hunting BL.755 CBUs, as deployed during the wars in Slovenia and Croatia during 1991-1992. (Artwork by Tom Cooper)

MiG-21M (L-15) serial number 22823 was delivered to Yugoslavia in January 1972 and initially served with 204th Fighter Aviation Regiment until re-assigned to 130th Squadron in July 1979. After undergoing overhaul – during which it was re-painted in aluminium overall and re-engined with the Tumansky R-13-300 engine (thus being brought up to the MF standard) – it was re-assigned to 130th Squadron of 83rd Fighter Aviation Regiment in Priština. In 1991, it had its tactical number crudely overpainted in grey, while later on it received a replacement ruder from another MiG-21, giving it this, rather 'patchy' appearance. In 1997, 22823 was converted to carry the LORAP reconnaissance pod, and assigned to 1st Detachment, 353rd Reconnaissance Aviation Squadron at Batajnica AB. This was the last operational airframe of this unique sub-variant, and the last operational reconnaissance MiG of the Yugoslav and then the Serbian air force. (Artwork by Tom Cooper)

MiG-21bis (L-17) serial number 17166 was assigned to 123rd Fighter Aviation Squadron of 83rd Fighter Aviation Regiment from its delivery to Yugoslavia in the mid-1980s, until the end of its career. It was badly damaged during Operation Allied Force in 1999 while still at Sjenica AB, and subsequently used as a decoy at Ponikve AB. As with all MiG-21bis of the RV i PVO, and later for most of the time with the Serbian Air Force, it was painted in air superiority grey overall. The jet is shown in markings of the SV as applied in 1992, and with a typical weapons configuration for interception operations as carried out by this variant: with a single R-13M on the inboard underwing pylon and an R-60MK on the outboard. (Artwork by Tom Cooper)

MiG-21UM (NL-16) serial number 16178 as seen while in service with 101st Fighter Squadron, 204th Brigade, Serbian Air Force, at Batajnica AB in 2015. Delivered to the RV i PVO in 1979, it served with nearly all of its fighter aviation squadrons over the following 20 years. Following the Dayton Agreement, it was disarmed in 1997 – which included the removal of its underwing pylons, all the weapons-related wiring and sights – and converted to a 'pure' supersonic trainer, designated the NL-16Š (š stood for 'školski' or 'trainer'). As such – and with its overall appearance in a rather sorry state – it was one of the last three two-seat MiG-21s still in service with the SV, as illustrated here. The inset shows the crest of the Moma Stanojlović Works. Constructed at Batajnica AB in the 1980s, this was responsible for the overhaul of all MiG-21s of the former RV i PVO and then the SV, from 1992. (Artwork by Tom Cooper)

MiG-21F-13 (L-12) serial number 22512, as seen shortly after its delivery, at Batajnica AB in 1963. Visible in the background are six light-yellow-overall painted F-86Es of the RV i PVO's aerobatic team. (Photo by M. Milenović)

MiG-21F-13 serial number 22510 soon after landing at Batajnica AB, while assigned to the 204th Fighter Aviation Regiment, in 1964. Notable is its highly polished 'silver grey' overall livery, consisting of two layers of clear lacquer mixed with 5% and 10% aluminium powder, respectively. (Authors' Collection)

MiG-21U-400 (NL-12) serial number 22911, as seen while serving with 128th Fighter Aviation Squadron at Batajnica AB, in the late 1960s. Painted in the same way as the MiG-21F-13s (some of panels appear darker because they were made of steel, which has bluish appearance), this jet was lost in a fatal crash on 14 September 1970. (MC Odbrana)

A pair of MiG-21PFMs (L-14s) of 204th Fighter Aviation Regiment seen at Batajnica AB in the mid-1970s. Both were photographed while attached to the APA-4M auxiliary power units installed on URAL-375D trucks. (Authors' Collection)

Three MiG-21Rs (L-14Is with tactical numbers 106, 104, and 107), and a single MiG-21PFM (tactical number 715), of 352nd Reconnaissance Aviation Squadron, seen in the 1970s. (Authors' Collection)

The MiG-21UM (NL-14M) serial number 22955 of 204th Fighter Aviation Regiment, seen during Exercise Golija '76. Notable is that the front part of the fuselage, nose of the drop tank, and wing-tips were all painted in light blue. Also notable is the armament consisting of Munja pods for 128mm unguided rockets, which were extensively tested during this exercise. (Authors' Collection)

A MiG-21M (L-15) serial number 22823 of the 204th Fighter Aviation Regiment, as seen during the Exercise Golija '76. Notable is the application of light blue colour on the front fuselage, nose of the drop tank and fin-tips and the armament consisting of Munja pods for unguided rockets. Launch rails for R-3S air-to-air missils were attached to the outboard pylons but RV i PVO's MiG-21s next to never flew with combined warloads during combat operations. (Authors' Collection)

This MiG-21M (L-15) – serial number 22811 – was photographed while serving with 130th Fighter Aviation Squadron of 83rd Fighter Aviation Regiment at Priština AB in 1979. It was one of the Yugoslav MiG-21Ms re-engined with more powerful Tumansky R-13-300 (MM-16) engines, and thus brought to the MF-standard. (Photo by R. Ivanović/MC Odbrana)

A MiG-21PFM (L-14) of 129th Fighter Aviation Squadron, 185th Fighter Aviation Regiment (Training), seen while landing at Pula AB in 1986. Notable is its aluminium overall colour (applied after an overhaul at the Zmaj Works), and the GP-9 gun-pod installed under the centreline. This jet was scrapped after a crash-landing in 1988. (MC Odbrana)

A trio of MiG-21bis/bis-Ks of 117th Fighter Aviation Regiment, seen while underway at high altitude in the late 1980s. The entire RV i PVO fleet was painted in air superiority grey overall, although some of the early examples had the last section of the fuselage before the afterburner left in steel-blue. (Photo by S. Šijačić)

After the withdrawal of the RV i PVO from Slovenia and Croatia in 1991, three ex-Iraqi MiG-21SMTs upgraded to the bis-standard (Izdeliye-50bis), were taken up by the RV i PVO and assigned to 126th Squadron of 204th Regiment. Between September 1991 and February 1992 they flew a number of combat sorties. This photograph shows two of them, serialled 21198 (rear/left) and 21168 (foreground): these were their original Iraqi Air Force serials. (Photo by M. Micevski)

A highly polished MiG-21R of 352nd Reconnaissance Aviation Squadron of 204th Aviation Brigade, seen at Batajnica AB in 1993. By this time, all the former Yugoslav insignia had been over-painted and replaced by the insignia of the new FR Yugoslavia, including roundels which were known as 'Pepsi-markings'. The port-side fuel tank was painted in air superiority grey, and thus had probably been taken from a MiG-21bis of 204th Brigade. (Photo by M. Micevski)

A close-up photograph showing details of the insignia applied on MiG-21s of 126th Squadron from the mid-1990s: a black and white lynx head atop a triangle/delta in light blue. (Photo by M. Micevski)

In October 2006, at the end of its overhaul by the Moma Stanojlović Works, this MiG-21bis received not only an entirely new camouflage pattern in two shades of grey, the application of which prompted significant discussion among the senior leadership of the air force, but also the insignia of the SV. Notable is the armament consisting of a total of six R-60 air-to-air missiles, and the 'last three' of its serial number repeated on the cover of its front landing gear. (Photo by S. Mihajlović)

This photograph shows MiG-21UM serial number 16185 during the last stage of an overhaul at the Moma Stanojlović Works in June 2016, when the jet received a two-tone camouflage pattern and a 'low-visibility' tactical number consisting of the 'last three' of its serial on the forward fuselage. (Photo by B. Dimitrijević)

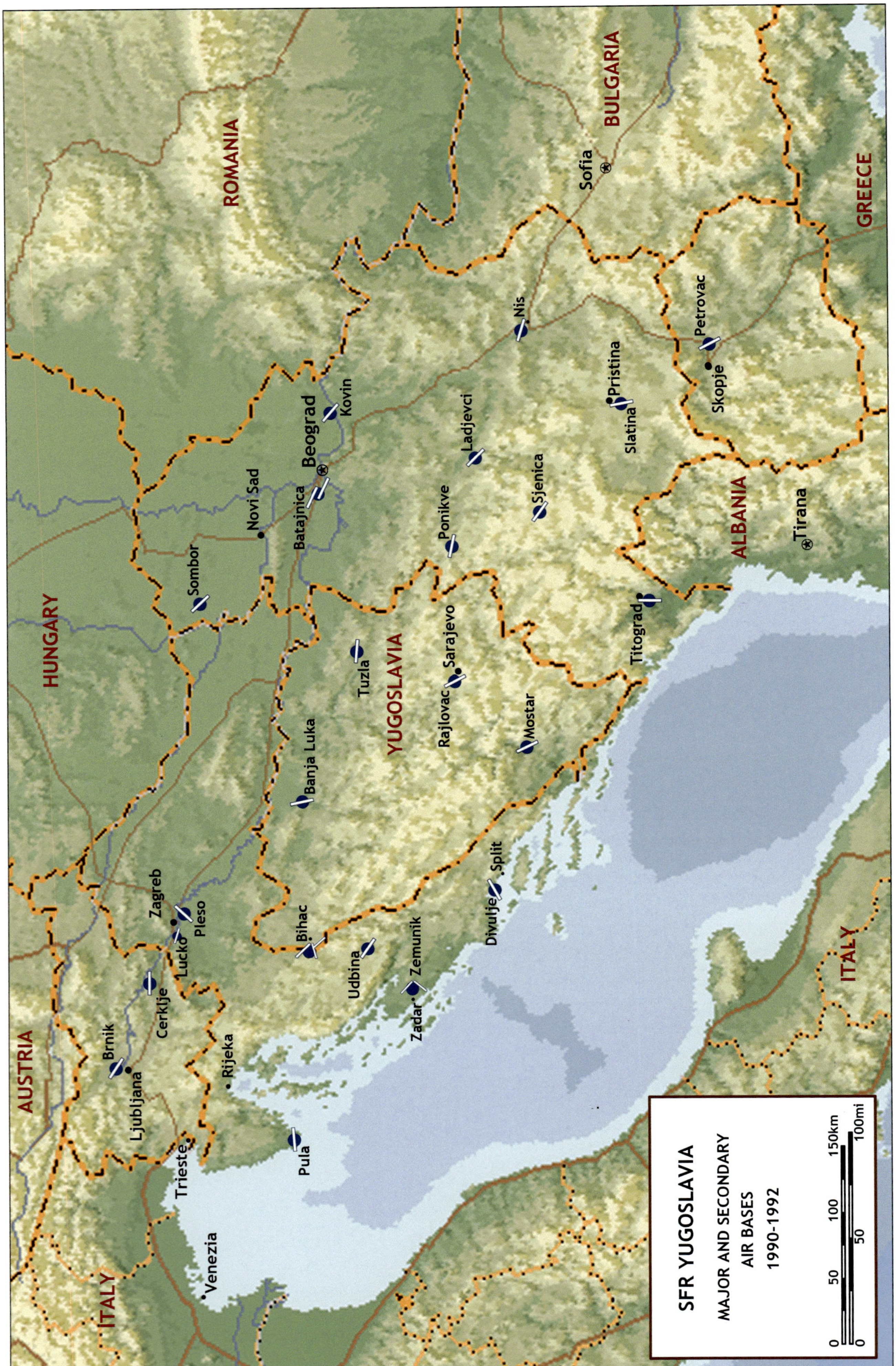
SFR YUGOSLAVIA
MAJOR AND SECONDARY
AIR BASES
1990-1992
0 50 100 150km
0 50 100mi
AUSTRIA
HUNGARY
ROMANIA
BULGARIA
YUGOSLAVIA
ALBANIA
GREECE
ITALY
ITALY
Sofia
Nis
Petrovac
Skopje
Pristina
Slatina
Kovin
Beograd
Batajnica
Novi Sad
Ladjevci
Sjenica
Ponikve
Sombor
Tirana
Titograd
Tuzla
Sarajevo
Rajlovac
Mostar
Banja Luka
Split
Divulje
Zagreb
Pleso
Lucko
Bihac
Udbina
Zemunik
Zadar
Cerklje
Brnik
Ljubljana
Rijeka
Trieste
Pula
Venezia

Moreover, in 1983, Soviet R-60 AAMs were introduced to the inventory and this type improved the interception capabilities of the Yugoslav MiG-21bis fleet. Previously, a team of Yugoslav airmen was sent to the Soviet Union to convert to the use and maintenance of this AAM and during 1983 training started in all of the MiG-21 regiments to use this type.[24] The R-60s were soon tested in launches at the Dugi Otok range over the Adriatic.

The introduction of the British BL.755 cluster bombs improved the strike capabilities of the MiG-21bis fleet. (Milan Micevski)

The period between 1982 and 1984 marked the start of wider training and usage of the MiGs "in auxiliary purpose" as the Yugoslav fighter authorities called strike missions: fighter sorties were treated as standard, while strike or fighter-bomber missions were regarded as "auxiliary". The last large exercise in Yugoslavia was "Unity 83" (Jedinstvo 83) at the Krivolak army range, in Macedonia, in September 1983. The main MiG-21bis units to participate were from 83rd and 204th Regiments. The preparations were carried out between 22 August and 14 September 1983 including some live gunnery practice over the southern Adriatic. No. 127 Squadron arrived from Batajnica and took part in the exercise from Skopski Petrovac and carried out strike missions with different strike ordnance including Hunting BL.755 cluster bombs, Munja unguided rockets and S-24 air to ground missiles.[25]

Strike ordnance used on the MiG-21bis: UB-16-57 57mm and M-74 Munja 128mm unguided rocket launchers, and PLAB-350 napalm bombs seen in front of No. 17220, at Batajnica in May 1988. (Milan Micevski)

Later, the commander of 83rd Regiment, Colonel Zoran Milićević fired two S-24 unguided missile in this exercise. As he recalls, at the moment of firing, the aircraft almost stopped in the air, and the pilot was forced to use the start (*zapusk*) button, since the launching of the S-24 actually pulled the air from around the aircraft, threatening to stop the engine. It was necessary to pull the MiG up to a higher altitude to avoid the turbulence caused by the explosion and the fragments of the rocket. "Nothing so ugly I haven't seen. But – very precise weapon" Milićević added. Lt. Colonel Nikola Ramljak, at that time a pilot with 83rd Regiment, was impressed by the effects of the BL.755 bombs and described them as "dreadful to watch from a side".[26]

At the same time as the Unity 83 exercise was carried out, General Inspection of National Defence organised a detailed check of 117th Regiment. Upon their orders, both squadrons carried strike missions on targets at Slunj firing range on 14 September. All kind of ordnance was used but noted as outstanding was a flight led by the Major Pejić which destroyed an AA position using the BL.755 cluster bombs. This regiment also undertook training in using the S-24 rockets, Munja 128 unguided rockets and PLAB-350 bombs.[27]

Introduction of the new weaponry significantly boosted the strike capability of the MiG-21bis fleet which more and more came to be regarded as a fighter-bomber rather than an interceptor platform. When the annual activities of all three regiments are analysed it can be seen that it was ordered that all fighter pilots "improve their skills in strike missions carrying out targeting, rocketing and bombing of the targets on the ground, using all available munitions, with the simple and complex aerial manoeuvres" with the MiG-21bis. No matter, the air defence planners still counted on the MiG-21bis being important in the broader system of air defence.

LORAP – American Reconnaissance Container Under the Soviet MiGs

As a part of the modernisation of the reconnaissance and intelligence assets of the RV i PVO, the American-built LORAP (long range aerial

Introduced in 1983-84, the Fairchild KA112 LORAP system boosted the aerial photo-reconnaissance capabilities of 352nd Reconnaissance Squadron. (Author's collection)

The MiG-21MF version was chosen as the LORAP carrier a since the type had enough underbelly clearance to accommodate the container. Here, No. 22867 can be seen during the initial trials at the Air Test Centre, in spring 1984. (VOC)

panoramic photographic) aerial photo-reconnaissance camera system was introduced in 1983. LORAP had been developed to meet the requirements for an operational system which could acquire reconnaissance intelligence from a long-range side-looking vantage point. The heart of the LORAP airborne system was a KA-112A panoramic camera. The photo-laboratory consisted of equipment for processing, deciphering and enlarging the images in three trailers and arrived in December 1982. A training course was organised at Moma Stanojlović Air Depot in Batajnica by the USAF and Fairchild technicians and two Yugoslav NCOs from No. 352 Squadron finished the course for this camera in summer 1983. On 4 September, a single KA-112 LORAP camera and two containers were delivered to the Air Testing Centre.[28]

It was soon discovered that MiG-21Rs of No. 352 Squadron were not able to carry the LORAP container on the under-belly pylon. After technical consultations and checks it was decided that the container should be carried by the MiG-21MF (L-15). This

A regular flying day in No. 352 Reconnaissance Squadron at Bihać in 1986 with two MiG-21Rs and a single US two-seater. (S. Janjanin/M. Vukojčić)

The last MiG-21 delivered to the Yugoslav RV i PVO was a UM in December 1986 and marked as 16185. It was assigned to 204th Fighter Regiment and is seen here at Batajnica half-a-year later. (Milan Micevski)

version had enough space beneath the fuselage to avoid the LORAP container touching the surface and the MF also had a more powerful engine than the R-version. The half-year trials started in the Aviation Test Centre in Batajnica in January 1984. In practice, the LORAP missions were not demanding, they just required maintaining a straight course, but the platform was at first unstable. The first MiG-21MF modified to the L-15M standard was transferred to No. 352 Squadron at Bihać on 15 June 1984. The converted MiG-21MFs were known as L-15M (modified) rather than L-15i (*izvidjač*, reconnaissance). They were stripped of radar and guns to enable better usage of the LORAP and to maintain proper flyability. A total of four MiG-21MFs (Nos. 22865 – 22868) were converted to LORAP carriers, while the L-14i continued to be used as platforms for Soviet reconnaissance pods.

On 1 August 1984, a photographic laboratory was set-up in the underground facility at Bihać and, finally, the system became operational. Officially it was known as the "L" container and in the beginning of its usage only four pilots were converted to fly the LORAP-equipped MiGs, later expanding this number to most of the pilots in No. 352 Squadron.

This system marked a significant step in equipping the Yugoslav aerial reconnaissance and it was used for wide and slope photography of the territories of the neighbouring countries. The camera did not have a horizontal limit and it could take images up to the end of the horizon. In practice, taken at an altitude of 10,000 metres it could cover up to some 110 kilometres of territory of Yugoslavia's neighbours. In practice however, weather conditions limited the effective usage of LORAP to some 50-60 kilometres.[29]

With the MiG-21R aging, the squadron maintained regular ELINT missions using the R containers, once per year, form Bihać, Priština and Titograd. Upon the estimations of the RV i PVO HQ they could be used several times for specific missions, but the estimations dated in 1987-1988 already marked R container as obsolete. In order to maintain flying proficiency among the pilots, No. 352 used two to four of the U/US/UM two-seater versions, and one or two PFMs. In this way the squadron preserved the resources of its reconnaissance MiGs.[30]

1986 – Last Reorganisation of the RV i PVO

The last peace-time reorganisation of the Yugoslav RV i PVO was ordered in February 1986. Based upon estimations of possible theatres of war and foreign aggression, Yugoslav territory was divided in three "theatres": central-north-eastern, south-eastern and north-western. Each of the "theatres" was covered by one Air Corps (Korpus RV i PVO) with HQs as follows: I at Belgrade, III at Niš and V in Zagreb. Each of the Air Corps were responsible for all of the RV i PVO units and bases in its territory. The Military Air Academy remained a separate peace-time entity. This reorganisation meant that two air defence divisions in which there were MiG-21 squadrons ceased to exist. Their assets were taken up by the Air Corps which now became responsible for all air missions within their area of responsibility.[31] Upon this reorganisation the nine MiG units were as follows:

- RV i PVO HQ: 352nd Reconnaissance Squadron – Bihać
- I Air Corps: 204th Fighter Regiment (Nos. 126, 127 and 128 Fighter Squadrons) – Batajnica
- III Air Corps: 83rd Fighter Regiment (Nos. 123 and 130 Fighter Squadrons) – Priština
- V Air Corps: 117th Fighter Regiment (Nos. 124 and 125 Fighter Squadrons) – Bihać
- Military Air Academy: 185th Fighter-Bomber Regiment – Training (No. 129 Fighter Squadron) – Pula
- Aviation Test Centre at Batajnica maintained a single MiG-21 for various trials

Yugoslav Air Defence authorities estimated in 1989 that fighter aviation operated a satisfactory number of assets to conduct its primary mission. The AAMs, avionics and especially MiG-21 radars had limited capabilities comparing to the "possible aggressors", however. The capacities for automatic guidance of the air surveillance network remained limited and had a direct effect on effectiveness of fighter aviation, which could be a potential problem in the event of war.

The available radars on the MiG-21 fleet in 1987 were: RP-21 (52 aircraft) with 20 kilometres range and RP-22SMA (82 aircraft) with 30 kilometres range. Both radars were noted as belonging to the "older generations without possibility of "looking down."[32] Besides 23mm guns in under-belly pods, Yugoslav MiGs possessed several types of AAMs with estimated ranges of between 0.2 to 15 kilometres. The older type R-3S and R-3R, and new R-60 could be used in VFR conditions and out of cloud day or night. The R-3R with radar guidance was usable in the clouds. Successful targeting was possible only approaching from the rear hemisphere of the target. The usual payload for the bis or earlier Ms were four air-to-air missiles. When the R-60 was obtained, a batch of 30 L-17s was converted to use APU-60II launchers extending its ability to carry a total of up to six of AAMs and though the Yugoslav air defence senior leadership considered that it was capable of carrying a total of eight R-60s, such combination was never officially documented. Generally, the Yugoslavs were aware of the "relatively modest" combat capabilities of the MiG-21s to maintain supremacy in their airspace, and their "inferiority" related to fighters *vis-a-vis* neighbouring air forces.[33]

This negative trend started to be overcome in late 1987 when a batch of 16 MiG-29s were purchased and delivered to No. 127 Fighter Squadron in Batajnica. This unit was in the conversion process during 1987-88, leaving 204th Regiment operational with Nos. 126 and 128 Squadrons.[34] Another MiG-29 squadron was not planned to enter service before 1995-2000. No matter the introduction of a squadron with the latest MiG fighter, for the bulk of the RV i PVO the MiG-

Table 5: MiG-21s of the RV i PVO, 31 October 1988[36]		
Regiment	Squadron	available aircrafts and versions
204	(126)	41 MiG-21bis, 4 MiG-21UM, 2 MiG-21U
117	(124, 125)	21 MiG-21bis, 2 MiG-21PFM, 2 MiG-21UM, 3 MiG-21U
	352.	10 MiG-21R, 4 MiG-21MF(LORAP), 1 MiG-21PFM, 2 MiG-21US
83	(123, 130)	13 MiG-21bis, 11 MiG-21M, 1 MiG-21UM, 1 MiG-21US, 2 MiG-21U
185	129.	15 MiG-21PFM, 7 MiG-21UM
Air Test Centre		1 MiG-21bis
Zmaj Air Depot		7 MiG-21bis, 4 MiG-21M, 5 MiG-21PFM, 6 MiG-21UM, 2 MiG-21US
Total		83 MiG-21bis, 19 MiG-21M/MF, 23 MiG-21PFM, 10 MiG-21R, 20 MiG-21UM, 5 MiG-21US, 7 MiG-21U

21 remained the dominant fighter and fighter-bomber platform, with estimations of a decline from 150-160 airframes to below 100 in 1994-1998.[35]

Generally in this period, the MiG-21 squadrons were trained to operate from different air bases, using civilian or so-called war-time "B" airports, with missions at the edge of their tactical radius. Such training was mostly remarked upon as "very good". Several large exercises were organised in this period such as Strike 88 and Power 88, with the largest air exercise being Deneb 89. MiG-21 squadrons were active participants in all of these exercises. By the middle of 1990, each of the squadrons had been trained for strike missions in up to four-ship formations. Compared with interception capabilities of the pilots, it was obvious that the MiG pilots' strike capabilities were growing, contrary to the basic fighter-interception capabilities which were of various levels.

MiG-21s at the Aviation Test Centre

The Yugoslav RV i PVO in its Cold War history showed a tendency to modify many of its aircraft and to adapt them to its needs. The MiG-21 was different, since most of the modifications were carried out by the Soviet producers. Most of the Yugoslav modifications, except for some minor ones, were limited to the introduction of different Yugoslav or Western ordnance, pods, tires or parts of the engines.

Since the MiG-21s were delivered to Yugoslavia without air-to-ground ordnance except for the UB unguided rocket launcher, most of the Yugoslav modifications were carried out in the domain of the ordnance. The first analyses and projects were carried out in late 1960s when the main idea was to integrate the already existing ordnance to the MiG-21 and to enable its use as a strike platform. The first task in 1969 was to integrate locally built HVAR 5-in unguided rockets on the MiG-21F. This was followed by the US 750lb napalm bomb trials held between September 1971 and June 1972.[37] In 1972, the L-128-04 unguided rocket launcher pod, later known as Munja, was integrated on MiG-21M (L-15) and later versions. Even modified obsolete German Second World War-era 250kg and 500kg bombs were tested. Soviet S-24 unguided air to surface missiles were tested in autumn 1976, followed by the British-built Hunting BL.755 cluster bomb in 1979. These were tested on the M-version, but later used with the MiG-21bis. Other trials included testing of some components and systems for use in future domestic projects, such as the fuel installation later to be used on the Yugoslav Soko Orao.

One unusual experiment was carried out in 1968. The idea was to use an F-86D Sabre Dog which had an autopilot as a target for air to air practice as the Yugoslavs estimated that live practice in the USSR was too expensive. The use of a LA-17 target drone cost US$95,000 as the Soviets explained to the surprised Yugoslavs after the

An unusual encounter at Pleso – Zagreb airport: MiG-21bis No. 17111 of 117th Regiment taxies near a brand-new Boeing 737-300 of Yugoslav Air Transport (JAT) in May 1986. (Milan Micevski)

Live missile practice over the Adriatic. MiG-21bis No. 17201 of 117th Regiment, seen during the launching of AAMs in October 1987. (Predrag Grandić)

Testing and usage of the US built HVAR 5-in rockets on the MIG-21F-13 was most likely the first integration of Western ordnance to the Soviet MiG-21. Testing was carried out between 14 February and 5 March 1969 in the Aviation Test Centre (VOC). Pilot Ante Sutalo carried out 15 test missions and approved this modification. HVAR 5-in rockets were used later on MiG-21Ms to a lesser extent. (Author's collection)

The first delivered MiG-21bis, marked as 17101, from December 1979 was assigned to the Air Test Centre at Batajnica, and was used for all kind of trials on this type until it crashed in November 1996. It is seen here during the DFA 001 target trials. (VOC)

first live practice in 1966![38] Aviation Test Centre (VOC) organised this trial over the south Adriatic. The Sabre Dog pilot Toma Djelošević was also a test-paratrooper and he took the aircraft up to altitude and jumped out of the open cockpit. In the meantime, pilot Ante Šutalo from 204th Regiment was cruising in a MiG-21F (22525) and when he received the signal fired an R-3S missile and claimed the empty fighter. The idea of using F-86Ds as target-drones was abandoned, however. Later, a much cheaper series of targets were developed. Different targets were tested with MiGs in the whole period including "Flag" (1982), Hayes (1983), GQ-2 (1983), and GQ-2YU (1989-90). Zmaj Air Depot also produced targets known as the SAM Z-80 family in 1990.[39]

The Air Test Centre in Batajnica operated a single MiG-21bis, No.17101, as the test bed for all kinds of testing since the beginning of the 1980s. For some wider or more demanding tests some MiGs were borrowed from the neighbouring 204th Regiment.

During the 1980s, different equipment and armament were tested on the MiG-21. Possibly the most important was the testing of the LORAP system on the MiG-21MF in 1983-84. Among the ordnance tested were FAB-100K and FAB-250K bombs in 1981 and 1982, aerial mines, the four-round Munja launcher for 128mm unguided rockets (both of domestic design and manufacture) in 1982, and PLAB-200 napalm bombs in 1988. The APU-60II launcher for two R-60s mounted on the MiG-21bis was first tested only in February 1992.[40]

A conversion of the MiG-21 (L-15) with a MM-16 engine was tested in 1982. Some engine spare parts and an extension of the MM-17 lifespan were tested in 1990-92. Different types of tyres were tested in 1988-89, and breaking parachutes in 1989-92. Finally, some avionics were tested such as the SO-1 prototype warning receiver in 1981, or APX 10V, 92V MAI10B, TPI 10 Teledyne systems in 1985.

Early 1989: Events in Kosovo

By the beginning of 1989, the security situation in the Serbian Autonomous Province of Kosovo became very fragile. An underground Albanian national movement had organised riots in 1968 and 1981 and in late February 1989, demonstrations again turned into full-scale riots that spread over many of Kosovo's municipalities. On 26 February, the Yugoslav Presidency ordered "special measures for keeping order and peace" in the province. The "measures" of placing the army at its highest readiness started on the following day at 06:00.[41]

83rd Regiment and 492nd Air Base were also alerted on the same evening of 26 February 1989. MiGs took off and made low-level flypasts over Priština and some other cities to warn the demonstrators and rioters. Despite some concerns over the security of their families back in Priština's suburbs, the MiG-21 pilots were eager to take some part in calming down the demonstrations. The low-level sorties were remembered by many. One of the pilots expressed: "It was pleasure to bump it!"

In mid-1990, upon orders for a general downsizing of the Yugoslav Army, some organisational changes occurred in RV i PVO and No. 128 Fighter Squadron in Batajnica was disbanded, passing its MiGs and

An excellent study of MiG-21M No. 22801, belonging to 83rd Regiment, seen in the late 1980s. After the overhaul in Zmaj Air Depot it was fitted with an R-13F-300 engine and brought to MF standard. (MRVA)

A three-ship formation consisting of 22952 (US/NL-14), 22803 and 22818 (M/L-15), of No. 130 Squadron of 83rd Regiment in the late 1980s. (Siniša Šijačić)

A MiG-21bis of 83rd Regiment, seen prior to entering the underground Objekt Rudnik at Priština Air Base in 1987. (83.lap)

"Flying day" in No. 130 Squadron. Three M/L-15s painted in air superiority grey being parked after a sortie. Among them 22824 and 22813 were fitted with R-13F-300 engines and brought to MF standard, while 22815 remained with its original R-11F2SK-300 engine. (Zoran Milićević)

During the 1980s it was the usual practice that fighter squadrons were deployed to different air bases for training. Here, one of the squadrons of the Bihać-based 117th Fighter Regiment is seen, with ground crew and an AN-26, is seeing during one such deployment. (Author's collection)

personnel to No. 126 Squadron. No. 352 Independent Reconnaissance Squadron in Bihać became part of 117th Regiment. [42]

Generally, the trend of the structural changes in 1990 was to downsize the number of conscripts in the aviation units, which led to a lack of the groundcrew in each squadron. Contrary to this trend, political and national tensions started to rise.

4
AT WAR

Table 6: Combat Sorties by 117th Fighter Aviation Regiment during the War in Slovenia, 27 June – 4 July 1991				
Date:	**Type of mission:**	**Carried out by Squadron**	**Number of sorties:**	**Total per day:**
27 June 1991	CAP	124. lae 125. lae 352. lae	6 6 6	18
28 June 1991	Combat sorties	124. lae 125. lae 352. lae	4 12 8	20+4
29 June 1991	Combat sorties	352. lae	4	4
30 June 1991	Combat sorties Weather reconnaissance	352. lae	3 1	4
1 July 1991	Combat sorties	124. lae 125. lae 352. lae	1 3 2	6
2 July 1991	Combat sorties	124. lae 125. lae 352. lae K-da 117. lap	4 6 4 8	22
3 July 1991	Weather reconnaissance	125. lae	1	1
4 July 1991	Weather reconnaissance	124. lae 125. lae 352. lae	1 1 2	2+2

The gradual dissolution of the SFRJ culminated in June 1991 when war erupted as the Federal government attempted to prevent Slovenia, and then Croatia, from establishing full independence. The RV i PVO units became involved in the clashes from the outset, primarily because of their importance and usage, but secondly because the air force had assets all over Yugoslavia.

Despite the fact that the rise of political tensions was clear to the whole of Yugoslav society, 1991 started for the RV i PVO as another 'regular' peace-time year. The MiG-21 units carried out their regular activities. By the beginning of 1991, the group of MiG-21 pilots that had finished the 39th Class of the Air Academy was deployed to 117th Regiment where they started the conversion to the L-17 version.

The spring brought the first clashes between Croatian Police and local Serbian armed groups and tensions rose. On 8 May 1991, V Air Corps HQ ordered increased combat readiness among its units based in central Slovenia, Croatia and western Bosnia and Herzegovina. The constant presence of 70% of personnel and retention of all of the conscripts in the units was ordered. In this period, 117th Regiment was ordered to establish detachments of four MiGs in 124th and 125th Squadrons and a pair in No. 352 Squadron tasked "for special purpose[s]."[1]

Following the unilateral decisions of Slovenia and Croatia to abandon the SFRJ, and upon the orders of the HQ RV i PVO, from 25 June 1991, all units were put on the highest state of readiness.[2] Although originally not intentioned to be run as such, combat operations began with the movement of ground forces towards the international borders of Slovenia on the morning of 27 June 1991. Thus began the war in Yugoslavia.

Short War in Slovenia

On the 26 June, the MiGs of 117th Regiment flew demonstration sorties in the zone around the Slovenian capital of Ljubljana in pairs at mid or low altitudes. For the first time pilots were issued with live ammunition for their side arms.[3]

The following day, army units rolled over the motorways and Slovenian countryside to the international borders with Italy, Austria

An impressive flight-line of the MiG-21PFMs belonging to 129th Squadron of 185th Regiment at Pula Air Base. The squadron, intended for cadet training, would soon be engaged in combat missions. (Miroslav Špica)

and Hungary. MiGs from Bihać and Pula were used in CAP missions, especially in so-called "closing" of the airspace around Ljubljana and Portorož airports. After Slovenian Territorial Defence shot down two RV i PVO helicopters and engaged in armed resistance on the first day of the clash, the RV i PVO started to use its combat aircraft from the morning of 28 June. MiGs from Bihać and Pula switched from CAP to CAS and interdiction sorties. At 14:45 that day, two MiGs fired unguided rockets at Dravograd to warn TO forces that had blocked an army unit on the road and targeted the massive barricades and road blocks erected by Slovenians along many of the important routes, as well as targeting communication and TV sites, airports and TO facilities in Kočevska Reka.[4]

A MiG-21bis of 117th Aviation Brigade returns to Bihać in 1991. The Munja launcher is still loaded with rockets, indicating that the pilot did not use them during the sortie. (Zoran Milovanović/MC Odbrana)

A pilot of 117th Aviation Brigade returns from a combat sortie in November 1991. His MiG-21bis No. 17210 carries only a UB-16-57 launcher beneath the wings. In front, No. 17213 is being parked on the ramp. (M.D. Ristić

The bulk of the MiG sorties were logged during the renewed clashes on 2 July. During the day MiGs targeted several TV, radio and communication sites at Nanos, Krvavec, Domžale, Boč and Kum and carried out several sorties at greater than the speed of sound over the Slovenian capital.[5]

During the strike missions MiGs from 117th Regiment used 57mm and 128mm unguided munitions, FAB-250 iron bombs and BL.755 cluster bombs, as well as S-24 rockets.[6] The fleet of MiG-21s that were engaged in combat consisted of three squadrons from Bihać and one from Pula, and none of them suffered any damage.

The clash in Slovenia was brought to an end on 7 July 1991 in accordance with a declaration signed in the Brioni Islands with the EC. A truce, the deployment of international monitors, a three month moratorium of secession of the republics of Slovenia and Croatia and the total withdrawal of the Yugoslav Army was agreed.[7]

The rear view of No. 17104 of 117th Aviation Brigade, damaged by a Croatian MANPAD during a combat sortie on 17 September 1991. (D. Buban)

The War Expands to Croatia

During July and August 1991, airmen originally from Slovenia and Croatia started to abandon the ranks of the RV i PVO and to leave for their home republics. By the end of July, 12 freshly promoted 2nd lieutenants (40th Class) who had earlier finished conversion on MiG-21s in Pula, were sent from the academy to 83rd Regiment in Priština which was noted by the RV i PVO as being the "safest place" where the young pilots could continue their training. This regiment maintained regular training and activities, and in August one squadron was prepared to deploy to Mostar, but the engagement was cancelled.[8]

During July and August 1991, following a rise in tensions in the areas in Croatia populated by Serbs, the Yugoslav Army, and particularly the RV i PVO remained at a high level of readiness. The RV i PVO conducted CAP and reconnaissance missions over Croatia and the RV i PVO did not engage again in the clashes in the second part of August. The MiGs returned to the scene and those from the Batajnica-based No. 126 Squadron operated over eastern Slavonia where the huge mass of the Army forces were engaged against Croatian forces.

Squadrons from the Bihać-based 117th Brigade became active flying over a wider area in Croatia, ranging from the Hungarian border to the Adriatic shores. In one mission on 26 August, a MiG was hit in the area between Sinj and Kijevo (Dalmatia) but the pilot managed to return safely.

Four days later, on 30 August, the QRA pair from Pula-based No. 129 Squadron took-off and intercepted a Boeing 707 cargo aircraft with Ugandan markings. The MiG pilots forced it to land at Pleso-

Captain Dronjak talks to the technical officer in front of 17205, armed with Munja launchers, in November 1991. This particular aircraft was hit by Croatian AA artillery on 26 August 1991 but after repairs returned to operational service. (M.D. Ristić)

Zagreb Air Base, where it was impounded by military police. The Boeing carried a cargo of weaponry for the Croatian forces obtained through illegal channels. The aircraft and cargo were detained and taken into the RV i PVO's inventory.[9] Two days later MiGs from the QRA at Bihać intercepted a DC-9 from Croatia Airlines and forced it to land in Split though this time, the aircraft was full of passengers.[10]

Following the ultimatum issued by the Croatian political leadership on 14 September 1991, many of the Yugoslav Army facilities were blocked by Croatian forces. This was also the case with Pula Air Base. No. 129 Fighter Squadron, which was stationed there, was ordered to abandon the base and to move inland. At 07:30 on 18 September, its six MiG-21PFMs and three UMs went to Bihać. The following day, two more PFMs arrived at Bihać. However, six MiG-21bis armed with AAMs were sent to Pula, to continue the CAP missions from the blockaded air base. No. 129 Squadron later continued to Tuzla Air Base in eastern Bosnia. A group of the cadets of the 41st Class, which started conversion on the MiG-21, was also in the squadron. No. 129 Squadron was then ordered to join 117th Aviation Brigade but it remained part of the parent 185th Regiment in Tuzla Air Base.[11] The detachment from Bihać remained in Pula, exchanging its pilots and aircraft until 22 October 1991 when the air base was finally abandoned and turned over to the Croatian authorities.

From 15 September 117th Brigade's MiGs overflew the Croatian capital of Zagreb as a warning to the Croatian authorities and population. On the same day, the MiGs from Bihać were engaged in attacks against airfields where Croat light aviation was believed to be hidden in order to prevent arms smuggling from Hungary, or the emergence of any kind of light combat aviation on the Croatian side.[12] Other missions were against Croatian forces at the front lines of Krajina: in the Banija, Kordun and Lika areas. It is difficult to identify all of the missions due to a lack of documents and the fact that pilots logged their combat missions as "navigation flight", "observing the communications" or simply with code "21/99".[13]

The first experiences from combat missions over Krajina and north Dalmatia led to improvements in tactics such as radio-silence during combat missions, a lower approach to Bihać home base, and the mounting of video cameras to record combat missions. The first experiences also led to harsh criticism and complaints by the pilots to higher authorities. The most common complaint of the pilots was the inappropriate usage of the MiGs in combat situations with tactical or even lesser importance. Chasing individual and isolated targets on the ground just compromised the general reputation of the Air Force as such.[14] Moreover the occasional truces that were negotiated and which led to sporadic ceasefires in certain areas just compromised the effects of the air strikes.

One of the youngest pilots of 117th Brigade, Second Lieutenant Roberto della Croce, wrote:

> In the beginning, only elder pilots went on to combat missions... At one moment we rebelled. We, young pilots. Why we sit aside? Why we are not taken on missions when we have completed the training in strike missions? The commander of the squadron warned us: these are serious business and there is no return when things started to move...[15]

The first losses to 117th Brigade occurred on 18 September when two MiG-21bis were lost. The first, No. 17212, was near Petrinja, and the other, No. 17109, was near Tounj. In the first case the pilot, Nuspahić, was killed, while in the second, the pilot, Mandić, was captured.[16]

By the beginning of October, the Bihać MiGs continued carrying out strike missions, mostly battle interdiction and strikes on important communication sites or bridges.[17] It used GSh-23 guns, PLAB napalm bombs and BL.755 cluster bombs. One mission proved particularly controversial, that against the Banski Dvori building in Zagreb old town – seat of the Croatian government – at around 15:00 on 7 October. According to a Croatian investigation, the attack was carried out by a pair of MiG-21s and a pair of Super Galebs, with the MiGs using 128mm Munja unguided rocket launchers. Regardless of what the aims were, the mission was unsuccessful and the material damage to the building did not prevent the Croatian parliament from confirming that republic's independence on the following day.[18]

Table 7: MiG-21s of 117th Aviation Brigade, 19 September 1991

	MiG-21bis	other versions	two-seaters	total
124 Squadron	13 (10 strike + 3 fighter)	-	-	13
125 Squadron	13 + 1 on trials (17213)	-	3	16
352 Squadron	-	MiG21R 5 + 2 unserviceable MiG-21MF 4 MiG-21PFM 1 + 1 unserviceable	MiG-21UM (NL14) 1 unserviceable	14

MiG-21R No. 26112, in which Captain Rudolf Perešin deserted from Bihać to Austria on 25 October 1991. It is seen here, in May 2019, in front of the Croatian Ministry of Defence, after Austrian authorities decided to pass this aircraft to Croatia. (Nikica Barić)

No. 17409 armed with 2 x 2 R-60 and R-3R AAMs, seen at QRA position at Batajnica. It sported an overpainted "squadron number" as a precaution against any observation from the enemy. (M. Micevski)

No. 352 Reconnaissance Squadron also became an important asset of the Yugoslav Army in fighting against Croatian forces. It carried out its tasks from the early days of the Slovenian conflict and the squadron was requested to monitor all of the withdrawals of the army forces from abandoned garrisons in Slovenia and later in Croatia. No matter its strategic role, the MiG-21 reconnaissance squadron started to receive requests from all around the front line for aerial reconnaissance. In the autumn 1991 operations, No. 352 Squadron carried out sorties sometimes at intervals of three hours. The requests were sometimes bizarre, where the commanders of the units asked for aerial photographs of just a mortar position, a fortified house, or similar tactical targets.

During the second half of October, No. 352 Squadron monitored the agreed withdrawal of the Yugoslav Army from garrisons in Slovenia towards the harbour at Koper, where they were loaded onto ships and sailed across the Adriatic to Montenegro. On the 25 October, a MiG-21R (No. 26112) with Capitan Rudolf Perešin was on an observation mission over Slovenia. After flying a couple of observation circuits he proceeded to Austria and landed at Klagenfurt airport. Perešin explained to the totally surprised Austrian officials that he left the Yugoslav Air Force because he was a Croat. It was quite a media sensation in Austria and elsewhere in Europe. It brought more attention to the conflict in Yugoslavia and served as an example for more Croat pilots to follow Perešin. The MiG remained in Austria, while the pilot joined the embryonic Croatian Air force.[19]

117th Brigade at Bihać continued to carry out combat sorties all over Croatia in late October and early November. For example, at the end of October MiG-21s carried out strikes on Croatian pontoon bridges at Kupa and against their tanks around Jastrebarsko on 1 November. Further strikes in mid-November were carried out at Grubišno Polje, against former army stores at Delnice, against pontoon bridges at Petrinja, against columns around Nova Gradiška, Našice, Sisak, Sinj, Otočac, and on the Kapavac communications centre at Papuk Mountain.[20] Combat operations were occasionally halted due to different temporary ceasefires. Moreover, the daily schedule of the brigade and its squadrons included some of the typical peacetime tasks: training of young pilots, training of older pilots in night or VFR conditions, some conversions, and even trials of the aircraft and equipment.

On 8 November, No. 352 Squadron lost another MiG-21R, No. 26109 equipped with a VHF system video camera, flown on a reconnaissance mission by former No. 352 Squadron commander Lt. Colonel Janjanin, now chief of the Intelligence Detachment V Air Corps. He managed to eject safely over the dense woods in Gorski Kotar but was captured by Croatian forces and sent into captivity.[21]

The Bihać-based squadrons continued to carry out strike missions from Grubišno Polje in the north over the Kordun and Lika area all the way to Vrsar airport in Istria using the standard assortment of strike ordnance. No. 352 Squadron also continued its regular aerial-photography missions. By order of 117th Brigade's CO, Lt Colonel Skadrak, in December a practice of "alternate functioning" was introduced between Nos. 124 and 125 Squadrons and the personnel had a chance to rest for 24 hours after a shift of 24 hours. This was a necessary step to overcome fatigue among ground crew and the lack of conscript personnel.

In the second half of December and up to 3 January 1992, 117th Brigade carried out strikes around Pakrac (16th), Karlovac (20th), Ramljani and Otocac (25-26th), Velebit Mountain (30 December and 2-3 January 1992) and stores near Rijeka (2 January). The last sorties were carried out on 3 January 1992. In this period, MiG-21bis No. 17130 was lost on 26 December, but its pilot Roberto della Croce managed to eject safely and was extracted by a rescue team led by

During a strike mission near the Slavonian town of Djakovo, MiG-21 No.17156 piloted by the Chief of Staff of 204th Regiment, Lieutenant Colonel Miroslav Vidaković, was shot down on 12 November. Vidaković was captured two days later when he attempted to cross the Sava River into Bosnia. Vidaković – a Croat himself – was later exchanged and returned to the Yugoslav side. (83.lap)

Three of the Iraqi MiG-21bis, Nos. 21168, 21198 and 21204, ferried from Zmaj Air Depot were used by No. 126 Squadron between September 1991 and February 1992. They had proper RV i PVO markings and serials painted on the Iraqi AF camouflage pattern. (M. Micevski)

Brigade CO Lieutenant Colonel Skadrak.[22]

Back at Batajnica, the MiGs of No. 126 Squadron had been carrying out CAP missions over eastern Slavonia since late August 1991. At the beginning of September, MiGs had been hunting Croatian AN-2s ferrying supplies and armament from Hungary into this area but failed to catch any due to the latter's low altitude flying. From 10 September, the MiGs joined the CAS mission in this area supporting army units in clashes all over eastern Slavonia. Some of the Jastrebs and Super Galebs were shot down by Croatian forces using MANPADs. Their pilots had been applying standard low-low attack tactics but RV i PVO HQ ordered in mid-September that all strikes should be carried out from altitudes above 2,000-2,500 metres. Strikes were flown daily, most of them at the request of units on the ground that were engaged in combat with Croatian forces. The CAS operations continued until the armistice at the beginning of January 1992.[23]

As a kind of replacement or back-up of the Bihać squadrons, No. 130 Fighter Squadron was sent from Priština to Bihać. It landed with ten MiG-21M and one two-seater. The idea in the RV i PVO HQ was that squadrons from the eastern part of Yugoslavia would relieve seasoned Bihać crews and enable pilots that were not engaged in combat to obtain first-hand experience. No. 130 Squadron mostly conducted QRA, CAP and escort missions over the Yugoslav Army units which were withdrawn from their surrounded garrisons in Croatia. Because of numerous ceasefires, the Priština MiGs were not engaged in too many CAS missions, if any. After 27 days spent in Bihać, No. 130 Squadron left on 26 December for their home-base at Priština. Harsh weather conditions turned their return into a drama. The first detachment hardly managed to land as it had used most of its fuel. Moreover, one of the MiGs had a slight incident and remained on the runaway, so the second detachment was ordered to proceed to Titograd-Podgorica Air Base and just managed to land with almost empty tanks. The third detachment, which waited to take off from Bihać, was ordered to remain there until the improvement of the weather in the morning.[24]

A distinctive feature of the Iraqi MiG-21s was the underbelly tropical filter. (M. Micevski)

On 26 November, the other squadron from 83rd, No. 123, deployed a four-ship detachment to Titograd with ten pilots. They established the QRA and air cover on the southern front and mostly conducted combat reconnaissance missions in the area between the Prevlaka peninsula and Dubrovnik, where heavy fighting raged between Yugoslav and Croatian forces. This detachment returned to Priština on 26 December.[25]

Table 8: MiG-21s of the RV i PVO, 30 November 1991

Regiment/Brigade	Squadron	available aircraft and versions
204	126	16 MiG-21bis, 4 MiG-21PFM, 2 MiG-21UM, 1 MiG-21U
117	(124, 125, 352)	28 MiG-21bis, 3 MiG-21PFM, 4 MiG-21MF(LORAP), 5 MiG-21R, 3 MiG-21UM, 2 MiG-21US, 2 MiG-21U
83	(123, 130)	18 MiG-21bis, 13 MiG-21M, 2 MiG-21UM, 2 MiG-21US, 3 MiG-21U
185	129.	9 MiG-21PFM, 10 MiG-21UM
Air Test Centre		1 MiG-21bis
Zmaj Air Depot		7 MiG-21bis, 2 MiG-21R, 1 MiG-21UM, 1 MiG-21US
Total		70 MiG-21bis, 17 MiG-21M/MF, 16 MiG-21PFM, 7 MiG-21R, 18 MiG-21UM, 5 MiG-21US, 6 MiG-21U

The rear section of No. 17402, belonging to No. 126 Squadron of 204th Fighter Regiment, seen here after its pilot Vladan Djukić "ran" into his own salvo of unguided rockets fired against targets in eastern Slavonia. He managed to return to Batajnica safely. (M. Micevski)

After long-lasting negotiations with the Croatian authorities, the Yugoslav Army managed to withdraw the Zmaj Air Depot from Velika Gorica near Zagreb in November 1991. Prior to the hostilities, there was a silent strike among workers who ceased to overhaul the airframes that were in the depot and the last overhauled aircraft rolled out in April 1991. During the summer of 1991 most of the flyable airframes were ferried to Batajnica Air Base. Some of them were airworthy but unfinished and were passed to Moma Stanojlovic Air Depot for finishing.[26] At least three Iraqi MiG-21s were ferried to Batajnica and added to No. 126 Fighter Squadron until February 1992.[27]

The Croatian ultimatum was that all of the withdrawal activities should be conducted by 100 workers in just 30 days. Even so, the RV i PVO managed to take out all of the 42 Yugoslav and Iraqi aircraft, 32 engines and 3,500 tons of other equipment, special machinery, instruments, spare parts, computers and documentation from the depot's premises. It enabled that the line for MiG-21 overhauls could be re-established in the Batajnica-based Moma Stanojlović Air Depot during 1992.[28]

After the truce in Croatia was signed on 2 January 1992, the clashes between Yugoslav Army and local Serbian Krajina forces on one side, and the Croatian force on the other side, ceased. A total ceasefire started on 3 January 1992 at 18:00. It was agreed that a UN peacekeeping force should enter the trouble zones and divide the factions and the RV i PVO halted its combat activities.

Effectiveness of the MiG-21bis in the 1991 Strike Missions

Analyses showed many shortcomings in the usage of the MiG-21 units in combat. Most of them were not related to the squadrons or the brigades/regiments which commanded them, but the criticism was of the upper echelons of the armed forces and ground forces HQs which requested their support in different types of tactical missions. It was estimated that a total of 75% of sorties were flown upon the request of ground forces in combat, while just 25% were planned by the Air Force. Other problems included the practice of the ground forces HQs asking for airstrikes during combat; urgently asking for the assistance over open communications and via several steps in the chain of command which compromised the secure planning of missions. It allowed the Croatian side in some cases to wait for the fighters with prepared AA artillery or MANPADs.

To improve the security in communications with ground forces and in the carrying out of strike missions, a new practice was introduced. The whole front-line was photographed with LORAP-system cameras. Photo-map "sections" were created and copied for each of the joint HQs with marked toponyms with random numbers or names to be used in the later communication and this improved the level of security when carrying out missions. Another practice was giving the details of the task to the pilots when they entered the cockpit, directly from the commanding officer.

Finally, there were difficulties in estimating the effect of strikes upon the enemy, which also influenced the type or the repeat of strike missions. The analyses dated in 7 January 1992, showed that No. 124 Squadron performed 220 sorties, No. 125 – 243, and No. 352 – 275.[29]

With 204th Regiment the percentage of combat sorties carried out "upon call" of the ground forces reached 80% of all performed missions. Half of this number qualified as "upon special request", which was not clearly defined as to its actual meaning. The MiGs operated in small groups, in most cases without any proper accurate intelligence on the situation on the ground. Direct control between commanders and pilots was established and it ensured safety to a certain extent. As

Beside unguided rockets, British-made BL.755 cluster bombs were mostly used in strike missions during the war in Croatia in the autumn-winter of 1991. Here, a technical service NCO loads the *"Be-eL-ovka"* onto No. 17227 belonging to No. 126 Squadron. (204.lap)

with 117th Brigade, it was difficult to estimate accuracy on targets. Both squadrons of the regiment (Nos. 126 and 127, the latter with MiG-29s) carried out a total of 568 missions between 25 August and 31 December. Among them were 476 sorties with usage of weaponry, while the others were CAP and other missions. In 147 sorties (26%) there was evidence of Croatian AA activity with AA guns or MANPADs. There were 16 instances of damage to the aircraft (13 to MiG-21s and 3 to MiG-29s). Among them 15 were categorised as "light", while only one was "heavy", or a loss (Vidaković). These results proved that MiG-21s were not just efficient but much safer than other RV i PVO strike types such as the Orao, Super Galeb or Jastreb.[30]

According to the 204th Regiment summary report, the MiG-21/29s used:

- bombs; BL.755 (170), FAB 100kg (110) and 250kg, PLAB 350 (102), 250kg (84), SAB 100 (10), KMGU (5) and BETAB 500 (4)
- unguided rockets; S-5K/M (3,499), Munja 128mm (816) and S-24 (8) and
- gun ammunition; 23mm (2,769) and 30mm (900).[31]

Total MiG-21 losses in 1991 were: four MiG-21bis and one "R" were claimed by Croatian AA fire, one "R" was "hijacked" and remained in Austria. In non-combat accidents three more bis, a single "R" and an "M" were lost. Making a total of 11 lost MiG-21s in 1991. Combat losses included one dead and three captured pilots.[32]

An Unexpected Incident: Shooting Down of EC Helicopters

Just when the RV i PVO units started to return to their regular "peace-time" activities, an unexpected incident occurred that spoiled the relations between the rump Yugoslavia and the European Community. At 13:42, on 7 January 1992, two unidentified "objects" were seen to enter Yugoslav airspace at the border with Hungary over the village of Kotoriba. The "objects" headed south. At 13:48, a MiG was alerted from the QRA pair at Bihać Air Base. After it was airborne and was heading towards the unidentified intruders, the pilot was ordered by the operations centre to shoot down both of them.

The pilot discovered the targets at 14:09 flying very low, maintaining an altitude between 100 and 200 meters. Without warning he fired several bursts from underbelly canon and then fired two R-60 missiles. One of them hit a helicopter which made a spiral dive to the ground and exploded; it was an AB-205A helicopter that belonged to the EC Monitor Mission which monitored the war and events in Croatia. The other helicopter, an AB-206A landed near the crash site. Four personnel from the Italian army aviation and one French naval officer were killed. The Yugoslav pilot made his way back to Bihać. Mission accomplished.

Prior to the incident, the helicopters were at Kaposvar in Hungary. The regional air traffic control in Belgrade warned the EC Monitor Mission that it had no control over the area where the helicopters intended to fly. The EC representative simply respond just saying "Yes, Yes!" On the next morning, when the helicopters took off, air traffic control in Budapest did not notify its colleagues in Belgrade of the take-off or estimated time when they would enter Yugoslav airspace. Moreover, the crew did not contact anybody when they left Hungarian airspace. Thus, the EC helicopters entered the zone which had earlier been proclaimed as restricted.[33]

The Croatian media were triumphant over this "crime with premeditation." A month later when a pilot deserted to the Croatian side, the name of the pilot who downed the EC helicopter was revealed as Captain Emir Šišić.[34] The incident was serious, and the Yugoslav authorities opened an investigation, suspended the RV i PVO commander General Jurjević and on the next day the Federal Minister of Defence, General Kadijević, resigned. Many years later Italy tried the pilot in absentia, and formally requested from the State Union of Serbia and Montenegro that Šišić must serve his sentence in Italy, which he did eleven years after the incident. No charges were

The damaged No. 17211 of No. 125 Squadron was used as a test bed at Bihać's apron during the autumn of 1991. It was later repaired and used in No. 124 Squadron between 1992-1993 at Ponikve, FR Yugoslavia. (M.D. Ristić)

brought against the senior officers who ordered Šišić to shoot down the EC helicopter.

Ceasefire Challenges

During the winter months of 1992, the MiG units returned to their daily routine, despite the feeling that a new phase of the war was expected.

On 4 February 1992, Captain I Class Danijel Borović of 124th Fighter Squadron took off from Bihać at 17:40 with MiG-21bis No. 17133 on a training mission. At 17:57 the ground control at his home base lost contact. What happened? Borović, who was Croat by origin, but remained in the RV i PVO and even carried out combat sorties in 1991, now decided to desert to the Croatian side. He simply proceeded north-west, to Pula airport and landed in the dusk. Croatia and its new air force received its first MiG-21.[35]

This was a serious blow to the RV i PVO, especially 117th Brigade. Suspicions against all pilots from the various Yugoslav republics now began to grow and their flying was restricted to two-seater MiGs or they were transferred out of the combat squadrons. The flying of single aircraft was banned on 7 February and three days later, it was ordered that two MiG-21bis armed with AAMs must be airborne while the other pilots had their training.[36]

The other, and much more serious problem that occurred in February 1992 was the deployment of UN peacekeeping forces. Bihać Air Base was on the border of Bosnia and Herzegovina and Croatia. On the Croatian side were the Serb populated areas which proclaimed themselves as Republic of Srpska Krajina. Now that Croatia became independent and the UN Security Council proclaimed Srpska Krajina as a UN Protected Area where, later, UNPROFOR forces were deployed. The arrival of the first UN monitors meant a restriction in flying from Bihać to only from the "Bosnian side" and not over UNPA Krajina/Croatian territory. This led to the introduction of new procedures of taking off and landing, as well carrying out daily missions.[37]

Despite the limitations, Bihać's MiGs continued to carry out regular air defence missions and even some exercises with the air surveillance battalion during February and March 1992. Just a couple of days prior to the escalation of the war in Bosnia and Herzegovina, it was ordered that only the QRA fighter pair should remain at the higher level of combat readiness![38]

MiG-21s in the First Phase of the Bosnian War

An explosion of clashes in many different areas of Bosnia and Herzegovina from 26-27 March 1992 were the first step in the upcoming Bosnian War and it was far from clear who was who on the ground. The exception was the advance of the Croatian regular forces in Posavina (middle-north Bosnia) and western Herzegovina. The sudden international recognition of Bosnia and Herzegovina on 6 April 1992 marked another important sign that the RV i PVO should abandon the country soon as possible.

The first engagement of Batajnica-based No. 126 Squadron's MiG-21s was on 28 March in the northern Bosnian region of Posavina. Two nights before, regular Croatian forces crossed the Sava river at Bosanski Brod and attacked the Serbian village of Sijekovac. They murdered a number of local inhabitants who did not manage to escape. Batajnica's MiG-21s, Super Galebs and Jastrebs attacked the positions of the Croatian forces in that area. They continued to do so in April, targeting the Croatian forces as they spread all over the Posavina area, and especially bridges and the pontoon crossing at the Sava River which linked Slavonski Brod on the Croatian side and Bosanski Brod on Bosnian side. In this period, the much slower Jastrebs were used occasionally as AFAC platforms marking targets for the MiG-21s which carried out the strikes.[39]

The bridge over the Sava was targeted in numerous sorties by No. 126 Squadron. It was an important objective since the Croatian regular forces used this bridge for deploying or supplying its forces in Bosnia and Herzegovina. The MiG pilots managed to damage the bridge, but without munitions which could destroy such concrete structures all efforts remained useless. In one such attack, Croatian units using MANPADs managed to shoot down MiG-21bis No. 17152 with Captain Slobodan Medic aboard. The aircraft crashed into the Sava near the public bathroom at Slavonski Brod. Since there was no trace of the pilot or his body, Medic was counted as missing in action for many years.[40]

352nd Reconnaissance Squadron MiG-21s (Nos. 26108 and 26107) seen after the abandoning of Bihać and joining 204th Aviation Brigade at Batajnica in May 1992. (M. Micevski)

The youngest fighter pilots of No. 129 Squadron of 185th Regiment seen during preparation for a combat sortie at Tuzla Air Base in the autumn of 1991. Their PFMs are armed with UB-16-57 launchers. (Aleksandar Kelić/MC Odbrana)

Table 9: Combat Sorties of RV i PVO MiG-21s in 1992[49]						
Unit	**Period**	**Total combat sorties**	**With use of ordnance**	**Reconnaissance sorties**	**Other combat sorties**	**Air bases used**
204th Av. Brigade (Nos. 126, 127 Sqns)	1 January – 31 May 1992	65	25	13	27	Batajnica and Bihać
83rd Fighter Regiment	4 – 11 May 1992	32	23	2	7	Bihać
No. 352 Reconnaissance Sqn	1 January – 31 May 1992	-	-	10	64	Bihać and Batajnica
NOTE: The authors have not discovered statistics for the disbanded 117th Aviation Brigade (Nos. 124 and 125 Fighter Squadrons) or for the disbanded No. 129 Fighter Squadron.						

The war in western Bosnia came later. The Bihać MiGs carried out their regular missions until the first week of April. After the regular Croatian forces moved from western Herzegovina, where they encountered Yugoslav Army and local Serbian forces near the of town Kupres and the surrounding highlands, the MiGs started to carry out observation, and later combat, missions over this area from 4 April.[41] The missions were carried out in the wider area of Kupres until 12 April when a MiG-21R (No. 26111) was shot down by the Croatian forces. Its pilot, Captain Grandić, ejected safely but fell directly into the hands of Croat forces and was captured.[42]

As the situation in the city of Bihać and surrounding areas started to complicate, the number of sorties by 117th Brigade started to decline. Local BIH territorial defence forces started to divide into Muslim and Serbian units, depending on the size of the population in a particular locality. After Bosnia and Herzegovina was recognised as independent, it was obvious that Bihać Air Base should be abandoned. The RV i PVO personnel mostly started to evacuate their families from the estates and to take care of their personal fate. There were few flying activities from the base after 17 April.[43]

Finally, on 21 April the word came: personnel were informed that Bihać Air Base should be abandoned, and the squadrons should move to Serbia. Evacuation started immediately. Some of the airmen decided to stay or to abandon the RV i PVO altogether. The three squadrons of 117th Aviation Brigade left Bihać in groups over three days from 22 to 24 April. A group of pilots from Batajnica arrived to fill the pilots' ranks and to enable efficient transfer of all of the MiGs to Serbia. The first to move was No. 352 Reconnaissance Squadron which arrived at Batajnica during 22 and 23 April and was placed under the command of 204th Aviation Brigade from the following day. The other two squadrons moved to the non-active air base at Ponikve in western Serbia. After their arrival at those empty facilities, No. 125 Squadron was disbanded. The remaining 124th Squadron was placed under the command of Priština-based 83rd Regiment. One of the most elite RV i PVO units, 117th Aviation Brigade, ceased to exist.[44]

But Bihać Air Base was not left without MiGs yet. On the morning of 25 April, a flight of MiGs from No. 126 Squadron at Batajnica arrived. They re-established a QRA fighter pair and prepared their MiGs to carry out strike missions. They were used against the infamous bridge at Slavonski/Bosanski Brod, but without success since their PRAB bombs were unable to destroy the bridge.

The Batajnica pilots were replaced on 4 May by the flight that arrived from Priština. They also took part in the strike missions besides their regular CAP missions. Their MiGs were also engaged in Posavina, around Bosanski Šamac, against the bridge at Brod and other locations where battle between Federal/Serbian forces on one side and Croatian/Muslim forces on the other, raged. Pilots remembered that due to a lack of food their daily ration mainly consisted of fresh trout but without bread, only with zwieback (a sweetened rusk or crispbread), for the entire deployment. Finally, on 11 May 1992 at 12:30, the last group of 83rd Fighter Regiment's MiGs took-off, carried out strikes in Posavina and continued to Serbia. It was the last sortie of the MiG fighters from Bihać Air Base.[45]

On the next day, 12 May, No. 129 Fighter Squadron of 185th Regiment left Tuzla Air Base, which had also been ordered to be abandoned and destroyed to prevent further use. With its ageing PFMs and UM two-seaters, the former Pula-based pilots and cadets landed at Priština Air Base.[46] The cadets (41st Class) were promoted earlier into 2nd lieutenants. During a small ceremony No. 129 Squadron was disbanded. The active personnel mostly left Priština, while "second lieutenants, eleven NCOs and all MiGs remained at rump" and were turned to 83rd Regiment, as the then Regiment CO Zoran Milićević recalls. Young pilots and those from the previous class of 1991 remained at Priština where they were ordered to continue conversion.[47] Another Yugoslav MiG-21 unit passed into history.

By mid-May, MiG-21s of No. 126 Squadron were being used to observe the withdrawal of the Yugoslav Army units from Bosnian garrisons. They made overflying sorties to secure the columns from possible attacks by Muslim (now BiH regular forces) or Croat forces. No matter that the last federal soldier had abandoned Bosnia and Herzegovina on 19 May, the MiGs of 204th Brigade carried out combat sorties over the Posavina area until the last was recorded on 31 May 1992.

On 15 May, further cases of desertion by Croatian pilots occurred. After the arrival at Ponikve, No. 124 Squadron organised regular training and flying. Since the unit was now in Serbia, there were no specific "nationality" limitations on the pilots. On that particular date, the training was organised in a larger formation. Suddenly Captain Ivica Ivandić, flying MiG-21bis (No.17167), abandoned the formation and continued to Herzegovina and the Adriatic Sea. After reaching the seashore he continued flying and landed at Split airport. Another pilot, 1st Class Captain Ivan Selak, aboard MiG-21bis (No.17235) abandoned his younger wingman, and continued over northern Bosnia directly to Pleso – Zagreb airport.[48] It was an enormous security blow for the RV i PVO (which now grounded all non-Serbian pilots), while a tremendous boost for the young Croatian Air Force which now operated three MiG-21bis fighters.

To summarise, during combat operations in 1992, the Yugoslav Air Force lost one MiG-21bis and one MiG-21R claimed by Croatian forces. One pilot was killed (although counted as MIA), while another one was captured. Two more pilots with their MiG-21bis deserted to Croatia.

Eleven MiGs of No. 352 Squadron seen on the central apron of Batajnica Air Base in late 1992. They sport new FR Yugoslavia roundels, widely known as "Pepsi", over the former Yugoslav "Red star-in-RAF roundels" markings used since 1944. Serials indicate that the line of consisted of basic L-14i and L-15M versions, with a single L-14 and NL-14 used for hack or training. (M. Micevski)

5
SHORTENED YUGOSLAVIA[1]

The Civil War of 1991-92 broke Tito's Yugoslavia apart. After Macedonia abandoned the federal state peacefully in February 1992, war moved to Bosnia and Herzegovina, which soon became recognised as an independent state by the international community. The new Federal Republic of Yugoslavia (FRY) was formed on 27 April 1992, consisting of two former Yugoslav republics: Serbia and Montenegro. The armed forces of the new state were named the Army of Yugoslavia (Vojska Jugoslavije, VJ), effective with 19 May 1992. The VJ consisted of all the assets that either remained in the territory of the two republics, or were evacuated there from Slovenia, Croatia, Bosnia and Herzegovina, and Montenegro. Unsurprisingly, the reformed air force consisted of units already stationed in Serbia and Montenegro and of those which made their withdrawal from the republics that had abandoned the SFRJ. During the late spring and summer of 1992, the air bases in FRY were overcrowded with aircraft, vehicles, technical equipment and above all: personnel that had become refugees in their own country. Some skilled personnel, originating from the now separated republics abandoned the new state and RV i PVO.

Much needed reorganisation started in September 1992. The RV i PVO was reorganised into the Air Corps (Vazduhoplovni korpus) and Air Defence Corps (Korpus PVO). The Air Corps controlled all aviation units, while the Air Defence Corps controlled all SAM Air Defence and Air Surveillance units.

There were five MiG-21 squadrons in the RV i PVO order of battle in late 1992. All of them were now part of two composite aviation brigades:

- 204th Aviation Brigade at Batajnica, with No. 126 Fighter Squadron (MiG-21bis), No. 127 Fighter Squadron (MiG-29), No. 252 FB Squadron (Super Galeb, Jastreb, Galeb) and No. 352 Reconnaissance Squadron (MiG-21R/MF)
- 83rd Aviation Brigade with HQ, Nos. 123 and 130 Fighter Squadron (MiG-21bis, M, PFM, UM) at Priština and No. 124 Fighter Squadron (MiG-21bis) at Ponikve Air Base.[2]

This organisation was actually an improvisation due to the war situation and experience. The brigades were a mix of existing fighter squadrons and those elements that had mostly arrived from Bihać and Pula/Tuzla in April and May 1992. At that time, the then leadership of the RV i PVO treated the MiG-21 fleet mostly as a strike/fighter-bomber force as it was used in 1991-1992, and uncertainty over the war and whether it would spread into the new Yugoslavia just strengthened this notion. Due to this way of thinking the MiG-21 fighter-interceptor role was neglected. This was especially the case with 204th Aviation Brigade which used MiG-21bis extensively in strike missions alongside the Yugoslav-made light strike Galeb G-4s and Jastrebs.

The further escalation of the war in Bosnia, resulted in all of the fighter units maintaining QRA assets at Batajnica, Priština, and the new air base at Ponikve from the summer of 1992. Moreover, the fighter detachment of No. 126 Squadron was deployed to the non-operational air base at Sjenica between late May and mid-August for the same purpose.

The new Yugoslavia was placed under a UN embargo in late May 1992, after the accusations that it had a military presence and

Reconnaissance MiG No. 26103 takes off, carrying a "D" reconnaissance pod, from Batajnica in 1993. (M. Micevski)

MiG-21bis, assigned to No. 123 Squadron of 83rd Aviation Brigade/Fighter Regiment at Priština during a training sortie. (via P. Grandić)

A MiG-21 overhaul line was set up in the Moma Stanojlović Air Depot. Here the technicians are working on No. 17232, sporting the primer green colour.

with strike missions in the event of war spreading to Serbia. This squadron, which remained after the disbanding of 117th Brigade, maintained regimental-strength by possessing 26 MiG-21bis and three two-seaters during late 1992. At the end of 1992, 83rd Brigade had an impressive total of 85 aircraft, including: 38 bis/L-17, 12 M/L-15, nine PFM/L-14 including two-seaters: 13 UM/NL-16, 3 U/NL-12 and 2 US/NL-14, and finally eight of the liaison Utva-66s.[3]

Generally, the large number of airframes that gathered at all fighter units enabled greater possibility for flying. However, proper maintenance started to be a problem for the technical service. The units waited for vital spare parts, such as engines, different regulators and wing parts much more than before 1991. Waiting time for some of the regular repairs took much longer, sometimes reaching 35 to 45 days. No. 124 Squadron at the new base at Ponikve had only basic squadron-level maintenance and nothing more. It was a serious problem since the squadron operated up to 30 airframes. The older versions, such as M (L-15) and PFM (L-14), had a serious decline in their airworthiness. Moreover, "nobody knew what to do" with the L-14 version.[4] To keep them airworthy, it was ordered that the elder pilots of 83rd Regiment carry out one sortie per month on each available PFM/L-14. Zoran Jakovljević who flew this type during the earliest days of his career, now carried out those flights. He compared this type with the current bis/L-17 saying "True violin... movable, light, potent!"[5]

In this period, maintenance and overhauling of the MiG-21s was established in the Moma Stanojlović Air Depot, at Batajnica Air Base. The technical line and numbers of disassembled MiGs were transferred in December 1991 from the Zagreb-based Zmaj Air Depot, which had overhauled MiG-21s since the late 1960s. Now all of the MiG-21 overhauling would be conducted by Moma Stanojlović.

Since most of the technical documentation of this Depot was destroyed in the air campaign of 1999, it is now difficult to estimate the success of "conquering" the overhauling process of the MiG-21s in Batajnica. But just comparing the numbers of finished airframes it may seem that this was a significantly slower process than at the more experienced Zmaj depot up to 1991. The other problem was Tumansky engine maintenance, which was carried out in the Orao Air Depot in Rajlovac, a suburb of Sarajevo. No matter that it remained in Serbian hands after May 1992, it was almost on the front lines and it

influence in the war that had spread in Bosnia. The embargo included the deploying of UN/EU monitors at each of the FRY air bases and these monitor teams controlled all of the RV i PVO activities up to late 1995. Moreover, NATO ships entered the south Adriatic establishing maritime control and the embargo. This caused the RV i PVO to make further deployments of its MiGs. At Podgorica Air Base (formerly Titograd up to 1992), a fighter QRA pair was established on 12 August 1992 to deal with any intrusion by NATO patrol or combat aircraft from the Adriatic Sea. The TDY shifts were provided by the crews and MiGs from both 83rd and 204th Brigades. In the same period, No. 352 Reconnaissance Squadron deployed its "R" and LORAP-equipped MiGs to Priština to carry out various ELINT and photoreconnaissance missions.

Meanwhile, 83rd Brigade turned its No. 123 Squadron at Priština into a training unit. All the former Pula-based L-14/NL-16s and almost 20 of the youngest fighter pilots were concentrated in this squadron. The other, No. 130 Squadron with its MiG-21M/bis was trained to carry out strike missions. They maintained QRA at Priština, as well as No. 124 Squadron at Ponikve. This ex-Bihać unit was also tasked

Table 10: MiG-21s of RV i PVO, March 1993

Brigade	Squadron	available aircraft and versions
204	126 352	18 MiG-21bis, 4 MiG-21UM, 3 MiG-21U 6 MiG-21R, 7 MiG-21PFM, 4 MiG-21MF(LORAP), 1 MiG-21UM
83	123 124 130	18 MiG-21bis, 8 MiG-21UM 20 MiG-21bis, 2 MiG-21US 12 MiG-21M, 9 MiG-21PFM, 5 MiG-21UM, 3 MiG-21U
Air Test Centre		1 MiG-21bis
Total (in service)		57 MiG-21bis, 12 MiG-21M, 4 MiG-21MF (LORAP), 6 MiG-21R, 16 MiG-21PFM, 18 MiG-21UM, 2 MiG-21US, 6 MiG-21U

Being prepared for a training sortie is No. 17104, which since 1992 had belonged to No. 126 Squadron. (Slobodan Mihajlović)

was hardly to be expected to show efficiency in overhauling the MiG engines. Problems in the maintenance of the MiG-21bis fleet start to be seen in late 1993 and in 1994, when the first decline in their airworthiness became obvious: from over 90 to 81% in 83rd Aviation Brigade.

Finally, the problem which emerged after two pilots – Croatians by their origin – deserted from Ponikve on 15 May 1992, led to a decision to ban flying for all pilots that were not residents of Serbia and Montenegro, or whose nationality "was not clear". No matter the fact that some of them logged sorties flying from Bihać in 1991-92, a practice called "forced vacation" was introduced until the situation became clear as to who really remained in the ranks of the shortened Yugoslavia's RV i PVO. This complicated daily routines, since some of the experienced MiG pilots were grounded, while others, some of them less experienced were burdened with missions, and it influenced morale. Both brigade HQs received the instructions that "no one should escape". It was a serious burden for organising regular flying and missions.

Training for young pilots that arrived from the academy continued now in Priština, introducing the new generations from 1993. Some pilots that had earlier been instructors in Pula were now sent on TDY to Priština. Some of the older pilots from 83rd Brigade, were promoted to instructors ("back-seaters") and managed to convert the youngest MiG-21 pilots with basic flying. It was a tough period. At the peak of the training, some of the older pilots logged up to 10-12 sorties per day, in many cases flying from both Priština and Ponikve air base in one day.

1994 – MiGs Return to Air Defence

By March 1994, a new organisation of the RV i PVO was ordered. Among many new solutions was returning the fighter aviation to Air Defence Corps. This change was in fact a return to pre-war concepts, where the MiG fighters were treated as part of the broader air defence system alongside SAM, radar and ELINT units. The combat experience of 1991-1992, where MiG-21s were treated as powerful strike platforms was now abandoned. This was the outcome of personnel changes at the top of the RV i PVO, as well as the emerging notion that NATO could carry out an air campaign against FR Yugoslavia.

In August 1994, two existing aviation brigades (Nos. 83 and 204) were returned to their pre-1990-91 organisation as fighter regiments with two fighter squadrons. No. 352 Reconnaissance Squadron was subordinated to the RV i PVO HQ, acting again as a strategic unit. This reorganisation was used to standardise 83rd Regiment completely on MiG-21bis/L-17. All of the obsolete versions were gathered in the squadron at Ponikve.[6]

Under this shift of versions, the squadron numbers were exchanged: No. 130 Fighter Squadron (earlier with MiG-21M/L-15) at Priština was now renumbered to No. 124 Fighter Squadron (MiG-21bis/L-17). No. 124 Fighter Squadron, at Ponikve, became No. 230 Fighter-bomber Squadron – the first digit "2" indicated that it had a FB/strike role. This unit was subordinated to 98th Aviation Brigade with its HQ at Ladjevci near Kraljevo, which mostly operated Yugoslav built Orao strike and reconnaissance aircraft.[7]

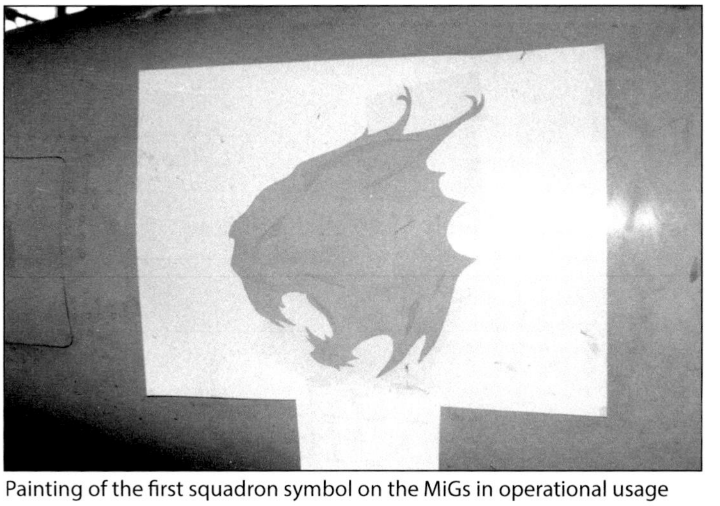
Painting of the first squadron symbol on the MiGs in operational usage in Yugoslavia. The head of a lynx was the symbol of No. 126 Squadron. (M. Micevski)

"The Demon" was briefly the symbol of No. 124 (ex-No. 130) Squadron in Priština. Few of the MiGs were painted with this symbol. It was soon discovered that some of the superstitious pilots used various excuses to refuse to fly in the marked aircraft. It was decided that the name of the squadron should be changed into "the Thunders". (Sasa Tirnanić)

A later symbol of No. 124 Squadron – "the Lion", seen painted on MiG-21UM No. 16158. (M. Micevski)

No. 126 Fighter Squadron in 1998: eleven MiG-21bis, a single UM and two more bis at the QRA position at the top of the photo. Though the scene may appear impressive, a much closer look reveals that most of the MiGs in the line were weary, some of them without serials or roundels on the fuselage. Only two of them were overhauled in Moma Stanojlović Air Depot: the tenth and last in the line. (M. Micevski)

Squadron Names and Symbols

Upon many different initiatives (including those of the authors) RV i PVO HQ in late 1994 approved that each of the squadrons could adopt a name and a badge. Engagement in combat during the war in 1991–1992 strengthened the notion and identity of being a member of a specific squadron among the personnel. Upon these initiatives each of the MiG-21 squadrons adopted a name and appropriate badge and markings.

Batajnica-based No. 126 Fighter Squadron adopted the title "Delta" and a lynx as its symbol, based on their call-signs. The badge consisted of a blue triangle, resembling the delta sign, with the head of a lynx, and this appeared on the noses of their MiGs. This symbol remained throughout the later history of the squadron, painted on the noses of the squadron's aircraft. One of the authors of this volume was engaged in the painting of the first symbol.

In the neighbouring No. 352 Reconnaissance Squadron, "the Stormer" (Oluj) was adopted as their name, along with a badge consisting of a photo-camera and the new name. It was inspired by a First World War Serbian Air Force Bleriot which carried the name of "Stormer". After this squadron was disbanded in 1997, the remaining 1st Detachment took-over the symbol of the parent No. 353 Reconnaissance Squadron "the Falcons" with split falcon wings on a blue shield. These symbols were not painted on their MiGs.

In 83rd Regiment the practice of marking the different squadrons dated back to spring 1989, when the CO of No. 123 Squadron, Major Urošević, introduced painting of the radome covers in yellow to distinguish his squadron from the other one which continued using the standard red-painted covers. Such practice remained until the end of the regiment's existence. In 1994, No. 123 Squadron adopted the name "the Lions", while No. 124 had chosen the name "the Demons", but soon changed to "the Thunders". 83rd Regiment did not practice painting its squadron badges on their MiGs, except for a few which sported the "Lion" symbol. In practice they still distinguish their MiGs by the colour of the radome-covers: red or yellow.

Decade of Technical Problems

The problem that emerged in 1994 and lasted almost to 2004 was lack of fuel. An embargo and emerging economic crises in FRY caused a drastic decline in flying hours per pilot. Until 1992, the usual annual average for the fighter pilots was around 80 flying hours. Now, fuel was delivered on weekly quotas, usually no more than 100 tons per combat squadron, which enabled a total of up to 100 flying hours for all pilots.

The number of pilots in the squadrons started to rise to 30 and more. This was the result of the number of MiG-21 pilots that had come from Bihać and Pula, as well as the annual arrival of the groups of young pilots from the academy. This caused lot of problems in organising the daily flying, training, pilot improvement and maintaining the reserve amount of fuel for use in case of emergency and war. Usually throughout the 1990s, the experienced MiG pilots managed to log

15-30 flying hours, while the younger pilots ranged between 5 to 10 flying hours annually, skipping most of the parts of the combat training. Those averages varied from year to year. Generally, up to 20 hours was the average that trained fighter pilots achieved annually during the in 1990s.

Dayton Accord downsizing in August 1996: a total of nine different MiG-21s were sent to the Yugoslav Air Force Museum at Belgrade International Airport. (M. Micevski)

Despite the variety of problems, larger air exercises started to be organised after 1994, and continued in 1996-1997. No. 126 and 230 Fighter-bomber Squadrons participated in some of these using their MiGs as strike platforms. The first aerial gunnery practice was held over the south Adriatic in September 1996, with usage of air-to-air missiles over ranges near the Montenegrin coastline and was the first live practice after many years. This was an opportunity for many younger pilots that arrived in the MiG-21 units between 1989 and 1991 to fire their first R-60 missiles, while a few older and middle-aged pilots used the cannon, against GQ-2 targets.[8]

No. 17409 was in the batch that was withdrawn to the museum, but in 1997 it was taken to the Air Depot, overhauled in 2003, and became the last MiG-21bis to be serviceable in 2015. (M. Micevski)

During the 1990s, the Aviation Test Centre continued to test different armament and equipment on the MiG-21bis, including OFAB-100 and 250K bombs (1992), KMGU-2 containers (1993), KVAB-04 training bombs, and KVB-4 and VAB-16 containers for training bombs, B-8M1 rocket launcher pods (1998) and HM-3/23 training bombs (1998-99). Then, different tests of KP-005 breaking parachutes (1994-1996), devices for firing 26mm IC-chaff (1996), devices for aerial gunnery including DFA/Q, GQ-2 and GQ-YU towed-targets (1995-19987), and different tyres (1997).[9] In November 1996, VOC lost its test-bed 17101 in a crash. Later, VOC obtained aircraft from 204th Regiment for further trials.

A LORAP container on a trailer waiting to be fitted under the belly of a modified MiG-21M (L-15M), prior to a reconnaissance sortie over Kosovo in 1998. (S. Mihajlović)

The "Dayton Accord" Downsizing in 1997

The end of the war in Bosnia was marked with the signing of a peace-accord in Paris in December 1995. Elements of the peace-treaty were set up later in several conferences with different subjects to organise peace among the former Yugoslav republics and nations. Under the sub-regional agreement signed in June 1996, limitations in quantities of combat effectives were introduced for FR Yugoslavia, Croatia, Bosnia and Herzegovina (including Republic of Srpska and Muslim-Croat Federation). Upon this agreement, FR Yugoslavia was allowed to possess 155 combat aircraft and 53 combat helicopters. All surplus aircraft were to be destroyed or if they remained in inventory were to be visually marked on the fuselage: disarmed/for research ("IR") or ground training ("GTA").

Upon this agreement, the FRY's MiG-21 fleet was significantly downsized in 1996-1997. The fleet of the older versions was mostly written-off from service. These included 15 PFMs (L-14), eight Ms and four MFs (L-15, L-15M), two-seaters: six Us and five US/UMs (NL-12, NL-14, NL-14M), 4 Rs (L-14i) and three later UM two-seaters (NL-16) and even two bis (L-17) making 47 airframes in total. The airframes were scrapped at Batajnica, Priština or Ponikve in front of the multinational verification commissions by cutting them at exactly marked positions. Moreover, seven UM (NL-16) two-seaters had their pylons for ordnance removed, and they were turned into unarmed trainers carrying the restyled marking NL-16Š, denoting *Školski* or training. The others with pylons were marked as NL-16Bs denoting their combat (*borbeni*) capability.

From a total of 65 different combat aircraft sent to the Yugoslav Aeronautical Museum at Belgrade International Airport, nine MiG-21s belonging to different versions were selected (1 L-14, 2 L-15, 2 L-17, 1 NL-12, 1 NL-14, 2 NL-16).

Upon this "Dayton" downsizing, as it was known among the RV i PVO personnel, two of the MiG squadrons were disbanded:

- No. 230 FB Squadron at Ponikve on 23 April 1997. This air base was abandoned, since it was less than 50 kilometres from the border between FRY and BiH. Some of the pilots and ground crew re-joined their former parent 83rd Regiment in Priština while others went to other units.
- No. 352 Reconnaissance Squadron at Batajnica, which now became 1st Detachment of No. 353 Reconnaissance Squadron on 29 May 1997. This squadron was an Orao-equipped unit, based in Ladjevci near Kraljevo. Since all of the LORAP-carrier MiGs were scrapped, a couple were chosen from among the L-15s and modified to L-15i/M standard to carry the LORAP container. Two "R" (L-14i) and three M (L-15) versions which had sufficient flying hours left remained in further service of this unit.[10]

6
KOSOVO CRISIS

Unsolved political and national problems in the south Serbian province of Kosovo and Metohija which had bubbled below the surface since the beginning of the 1990s, had by early 1998 turned into open clashes. The Albanian national movement led by the younger generation of leaders abandoned idea of remaining in the shortened Yugoslavia and Serbia. They started with guerrilla actions and quickly during 1998 spread into a broad movement which managed to seize control over portions of Kosovo. It was a sign that Yugoslav Army units would be deployed from their garrisons in Kosovo's cities.

Widespread clashes in the summer of 1998, and massive usage of FRY/Serbian security forces led to a reaction in the international community. NATO soon stood against such massive usage of power against the guerrillas and, as often added, the civilian population. Political and media pressure was added to the growing military presence of NATO air assets in bases in Italy and the conduct of several air exercises over the Balkans. As the political negotiations with Belgrade faltered in the first half

A QRA pair of No. 126 Squadron seen on a rainy day in 1998. Nearest to the camera is No. 17401, armed with two R-60 and two R-3R AAMs. It would be destroyed on the ground during Operation Allied Force in May 1999. (M. Micevski)

Passing a line of SA-341/342 Gazelles, No. 17221 taxies out to the runway of Batajnica Air Base in May 1998. It was overhauled in Moma Stanojlović between 1992 and 1994, as indicated by a small Air Depot sign on the fuselage. (M. Micevski)

Seen over Kosovo is MiG-21UM No. 16155 of 83rd Fighter Regiment. (Siniša Šijačić)

No. 17155 served with 83rd Regiment for most of its career. Prior to the war with NATO, it was transferred to No. 126 Squadron in Batajnica. It survived Operation Allied Force. (Siniša Šijačić)

Seen after crash landing in 1998 is No. 17159. It was a MiG-21bis which spent nearly its entire career assigned to No. 123 Squadron of 83rd Fighter Regiment at Priština. (HQ RV i PVO)

of October 1998, the deployed NATO air assets expected the signal to start attacks on FR Yugoslavia.

RV i PVO units were gradually activated during the beginning of October and both fighter regiments were alerted. In 204th Regiment, only MiG-29s were engaged, while the MiG-21s were stored in HAS or dispersed all over Batajnica Air Base. In 83rd Regiment, some of the MiGs were moved into the Rudnik underground objekt. A few months earlier, the pilots of this unit had started to fly with their sidearms loaded. Security at the aprons and ramps of Priština Air Base was enforced by armed guards and the pilots lived and carried out their missions in near-to-war conditions long before the crisis started.

This sudden crisis was resolved with FRY President Milosevic's accord with US envoy Richard Holbrooke on 13 October 1998. The "Activation Order" for NATO air strikes was suspended but not cancelled. On the next day, RV i PVO HQ ordered the return to regular activities. The crisis began to calm down, no matter that in Kosovo the clashes between guerrillas and Yugoslav/Serbian forces continued.

It was agreed that NATO should carry out reconnaissance flights to monitor the situation in the province. The RV i PVO fighter aviation was not allowed to be airborne when NATO (USAF) reconnaissance flights were carried out and monitors were exchanged between RV i PVO HQ in Zemun and NATO CAOC in Vicenza.

On the first day of air monitoring over Kosovo, Captain Roberto della Croce was training in IFR flying with one of the younger pilots. As they started to descend through the clouds, della Croce noticed a large shadow passing over his cockpit. "I look above and realise that some 200 metres above is Predator UAV. I would gladly burst it away with my afterburner. But except for showing it the middle finger nothing was allowed to be done. I hope that they managed to photograph it. We carried on with our business."[1]

The operational capabilities at the end of 1998 were improved, but far from suitable for real combat operations. The figures from 204th Regiment showed average flying hours per pilot were just 25. Training of small groups of the youngest pilots was conducted as and when it was possible. Live practice using the R-3S and R-60 was carried out over the south Adriatic using PRM-200 air launched targets. Average airworthiness of the L-17 fleet was 55% and for the two-seater NL-16s only 30%. The problem with airworthiness of the two-seater fleet was caused by the "slow dynamics of overhauling."[2]

Flying in 83rd Regiment was intensive during 1998 and even some of the younger pilots managed to log up to 30 flying hours in that year. Flying was organised even if it was a "technical day" or "reserve day." During the intensive flying in this period, the regiment lost three MiG-21s but there were no fatalities.

The two-seater UM No. 16178 seen over Kosovo in the mid-1990s. Prior to Operation Allied Force it would be detached to 1/353rd Detachment. It survived the campaign and remained one of the last UM airframes to be operated as late as 2019. (Siniša Šijačić)

A total of eight MiG-21s were in 1/353rd Detachment: four M, two Rs and two UMs. A single bis was used by the Aviation Test Centre, and the few remaining MiG-21bis/UMs were in Moma Stanojlović Air Depot waiting for overhauls.

Gradual Mobilisation

By the beginning of 1999, the situation in Kosovo and in the international political arena had failed to reach a peaceful solution. War gradually seemed to be inevitable.

In late January 1999, the RV i PVO senior leadership started a plan named "Air Defence Operation" (*Operacija PVO*). It was planned that MiG-21 fighter squadrons should be engaged in the air defence of important/strategic targets, but the main role was given to the SAM units. Surprisingly, RV i PVO planners expected NATO airborne landings in Kosovo and 83rd Regiment was tasked to engage the enemy airborne formations!

In a later plan, issued on 15 March 1999, it was ordered that the Air Defence Corps should prepare a flight of eight MiG-21s to carry out strike missions against NATO forces in Macedonia, deployed in the area of the Krivolak range. It was added that the mission should be carried out only if President Milosevic approved it. No matter the planning, the Commander of the RV i PVO, General Smiljanić, warned President Milosevic on its inferiority in comparison to the NATO forces in case of attack. Smiljanić, a former MiG-21 pilot, especially noted the inferiority of this type facing the aggressors.

The gradual mobilisation of RV i PVO units started on 16 February 1999. Two days later, personnel started to disperse the ammunition, spare parts, fuel and other equipment from their premises to locations around the air bases. On 20 February, QRA fighter pairs were doubled and now there were four fighters in QRA at Batajnica and Priština. Four more MiGs were in higher alert status as back-up in both regiments. In the meantime, the technicians of the Moma Stanojlović Air Depot together with technicians from the fighter units managed to recover the MiG-21 fleet in the squadrons to 98% airworthiness (comparing favourably to 71% in the MiG 29 squadron).

According to the author's research, prior to the air campaign there were a total of 72 available MiG-21s. Those were 51 bis, 15 UM, four M and two R. On 23 March 1999, RV i PVO had following fighters at its disposal:[3]

- 83rd Regiment: 30 MiG-21bis and 6 MiG-21UM with a total of 48 pilots
- 204th Regiment: 16 Mig-21bis and 3 Mig-21UM (two-seater No. 16182 was out of service)
- 16 MiG-29 and 2 MiG-29U (with six and one respectively out of order) with a total of 52 pilots[4]

Deployment of the Yugoslav MiGs prior to the Air Campaign

The talks in Belgrade between Milosevic and Holbrooke failed on 22 March 1999 and was a sign that air strikes would commence. The Secretary General of NATO, Xavier Solana, approved the launch of the campaign called Operation Allied Force. 23 and 24 March were used by both sides to take up positions. Yugoslavia declared near-to-war status at 20:00 on the evening of 23 March.

At 22:15 the combat alert was given to RV i PVO. Personnel started to gather in their units. At 23:45 an "Air Alert" was given to 204th Regiment, and at 00:00 to 83rd Regiment. Upon this signal, the ground crew of No. 126 Squadron started to move their MiG-21s out of Batajnica Air Base dispersing them alongside the nearby Belgrade-Novi Sad motorway. They were returned back to base in the early morning of 24 March. Its MiG-21s were dispersed during the following morning at several gates and aprons of Batajnica Air Base, in groups numbering three to six aircraft.

On the morning of 24 March, it was ordered that both 83rd and 204th Regiments should deploy some of their fighters out from their home bases. 204th deployed two MiG-29s to Niš and Ponikve, and one MiG-29 to Podgorica. 83rd Regiment deployed four MiG-21s to Sjenica, two each at Podgorica, Niš and Ponikve. These flights were carried out in radio-silence and at low level, first to Podgorica at 14:25, and last to Niš at 14:55. Previously on the same morning, the ground crew teams were deployed by AN-26 to the air bases to await the MiGs. Each group had technicians sufficient to solve I and II degree maintenance issues.

Besides Batajnica and Priština, the RV i PVO established mixed fighter detachments from both fighter regiments in four other air bases: at Niš: two MiG-29s and two MiG-21s; at Ponikve: two MiG-29s plus two MiG-21s; at Podgorica: one MiG-29 plus two MiG-21s; and at Sjenica: four MiG-21s.

83rd Regiment's MiG-21s were armed in basic fighter configuration with AAMs. None of the deployed groups was given any particular task. They were ordered to report to the most senior commander at the air base and to wait further orders. The detachments were commanded by the MiG-29 pilots from 204th Regiment who were elder and more experienced than 83rd Regiment's MiG-21 pilots who were mostly younger pilots, captains that had finished the academy in 1989-1993. Some of the MiG-21 pilots had combat experience from 1991-92, while others only had live firing experience from training. Some of the pilots had never previously landed at the air bases where they were now deployed. Contrary to the MiG-29 teams, which came with replacement pilots, there were none available for the MiG-21s.

On the afternoon of 24 March 1999, all of the deployed MiG detachments were at the highest state of combat readiness, waiting to get airborne and intercept the NATO strike packages.

Fully armed and attached to the APUs (APA) are Nos. 17228 and 17401 of No. 126 Squadron at the QRA station at Batajnica. Both would be destroyed on the ground during Operation Allied Force. (M. Micevski)

Capitan Oluić ("Stormy") of 83rd Regiment was the MiG-21 pilot who came closest to taking off on the night of 24 March 1999 to intercept the NATO strike packages. At 20:40 at Ponikve, the signal was given "All in the air!" The CO of No. 127 Squadron, Lieutenant Colonel Djukanović, ordered that two MiG-29s and one MiG-21 should take off. First to enter the runway was Major Milutinović, followed by Djukanović on the MiG 29s. The third was a MiG-21bis with Captain Oluic. Milutinović took off at 20:45. Djukanović's ground crew was a bit slower and they prevented Oluić's MiG-21 from passing to the runway. Major Milutinović had already taken off; when the others proceeded to the runway, they were halted by 31st Operations Centre in Kraljevo who commanded the alert. The order came that no MiG-21 should take off as it was estimated that they would be wiped from the sky by much more potent NATO formations. (MC Odbrana)

Allied Force: the MiG-21 fleet order not to take off!

The air alert signal was given at 18:40 on 24 March 1999. Ten minutes later almost all of 83rd Regiment's personnel were in the underground Objekt Rudnik. In Priština, 20 MiG-21bis and six MiG-21UMs remained. Except for the four MiGs in QRA, all others were hidden in the underground objekt. At 19:30, 280th ELINT Centre registered the first USAF strike package in Albanian airspace and 11 minutes later the first cruise missile launched from ships in the Adriatic hit selected target in FR Yugoslavia. The last European war in the 20th century had started.

At the same time as the first air attacks were carried out, the Chief of the Staff Air Defence Corps General Grujin ordered that eight of the MiGs should be taken out from the underground objekt and deployed outside around the base. It could be that the senior leadership feared that air strikes may hit the entrances to the objekts and prevent the fighters from being engaged in further combat. Nearly two hours later Tomahawk cruise missiles missed the entrances at the sides of the objekt. The strikes did cause the south entrance to be blocked by the masses of earth and stones which tumbled down, though this was cleared during following days with bulldozers.

Pilots in QRA were sitting in their cockpits, headed by Colonel Urošević, CO of 83rd Regiment, but the order to take off was not issued. In other air bases, after the alert was given, the MiG-29s took off to face the NATO formations.

In the moment that the campaign started, General Smiljanić banned the usage of the MiG-21s in the interception of the NATO formations. In his memoirs he stated that most of the RV i PVO personnel understood this decision. Former RV i PVO CO General Veličković, a MiG-21/29 pilot and ardent opponent of NATO, who

83rd Regiment maintained the QRA with its MiG-21s. Here, the QRA is station seen immediately after one of the NATO strikes. (M. Špica)

Reconnaissance Detachment 1/353rd lost three of its MiG-21Ms during the early days of the campaign. Here the wreckage of No. 22816 is seen in the scrapyard of Batajnica Air Base. (S. Mihajlović)

was on duty as the assistant to the Minster of Defence, was against this decision and openly disagreed in the top political and military circles. Finally, President Milosevic summoned General Smiljanić on 15 April asking him to explain his decision. Smiljanić explained that most likely no MiG-21 which faced the NATO strike packages would be able to avoid being shot down. He added that such result would be a tremendous propaganda boost for NATO. Milosevic listened carefully, agreed with Smiljanić's decision and closed this question. The MiG-21s remained grounded. General Veličković continued to dispute this decision, until 4 May when the CO of 204th Regiment, Colonel Milenko Pavlovic, flying a MiG-29 was claimed by a F-16CJ of No. 78 EFS USAF over Valjevo and killed. Veličković approached Smiljanić and apologised for insisting on the usage of MiG-21s against the NATO forces.

Despite Smiljanić's decision, the MiG-21s were kept in QRA in Batajnica (two along with two MiG-29s) and Priština. Whilst they were sometimes taken up to readiness No. 1 the interception sorties were not ordered. The QRAs and high readiness were continued through April with MiG-29s. Occasionally up to four MiG-21s were brought up to the highest combat readiness, such on 8 April in Batajnica. Most of the activity was actually just hiding and waiting.

When Kosovo Albanian guerrillas squeezed the border post of Košare by attacking it from Albanian territory, it was ordered on 12 April that 83rd Regiment should use four MiGs with BL.755 cluster bombs to help the border unit which was under heavy attack. On that evening, Colonel Urosevic choose three older pilots for the mission that he would lead. General Smiljanić ordered the mission at 18:45. The armourers removed the AAMs from four MiGs which were outside of the Rudnik underground objekt, however, the delivery of the BL.755 cluster bombs was late, and at 19:31 the general cancelled the mission. The idea of conducting such strikes was abandoned on the following morning.

Hammering the MiG Bases

Back at Batajnica Air Base on the evening of 24 March, the first NATO strike occurred around 20:40. Precision Guided Munitions started to hit the HASs and other positions in the base. The attacks continued until 04:00 on the morning of 25 March. One MiG-21 was destroyed in a HAS, another two were damaged at the apron. The complete stores of ammunition and spare missiles of No. 126 Squadron were destroyed in another HAS.[5] On the following day, 25 March, in a HAS of 1/353rd Squadron, a further two MiGs were destroyed. In the following days the hangar of 1/353rd was destroyed, causing the destruction of nearly all of the unit's MiGs and piston-engined Utva trainers which were inside.

It was a bitter lesson that Cold War-era Hardened Aircraft Shelters are useless when attacked with precision munitions. Even worse, a hit on a HAS meant the aircraft's complete destruction, which was not the case if the aircraft was targeted outside on the apron or a taxiway. This system of protection proved useless and it was concluded that the aircraft on the ground should be kept from destruction by using other methods of hiding or improvisation. From 26 March the ground crew at Batajnica started to move MiGs all over the air base which reduced losses.

Wreckage of a MiG which was damaged while hidden on the Belgrade-Novi Sad motorway near Batajnica Air Base. (204.lap)

The heavily-damaged two-seater 16151 remained in "beyond repair" condition and was scrapped many years after the campaign. (204.lap)

Hardened shelters did not provide protection for the Yugoslav Air Force from PGMs. One of 83rd Regiment's MiGs remained in this HAS at Ponikve and it was destroyed along with it. (S. Mihajlović)

Hidden amongst the trees, No. 17207 was discovered and destroyed on 29 May 1999 by a USAF strike package. (204.lap)

At Priština, beside the infrastructure, NATO strikes on the underground Objekt Rudnik continued. Soon it was discovered that it was psychologically difficult to remain in the underground objekt under the air attacks which continued every night. Vibration, clouds of dust and earth, fear of being crushed inside the objekt, meant that on 1 April the order came that all the personnel should abandon the site and move "out of the fence" in the vicinity of the air base.

Entrances to the underground objekt were especially targeted and on one night in April missiles hit the protective cover and damaged the northern entrance into the objekt. General Smiljanić inspected the site and approved the idea of 83rd Regiment's commander Colonel Urošević that aircraft, which were in the underground objekt were to remain, no matter the damage to this entrance. Underground Objekt Rudnik proved its value. It had suffered 17 air strikes in total, mostly with the direct hits, but remained operational. All of the MiG-21s which were inside the objekt remained intact and airworthy. The MiG-21s outside the underground objekt were soon to become targets for the NATO strike packages. In total, six MiGs were destroyed in Priština Air Base.

At the end of April, 83rd Regiment took advantage of the short break in the daily strikes and successfully managed to relocate six MiG-21s from Priština. Four of them landed at Sjenica while two landed at Niš Air Base. Those sorties were carried out during the break in the NATO air strikes, flying at low level and avoiding the NATO fighters which were on CAP missions. This deployment was made with the idea to disperse the MiGs to other bases. Soon, most of them became targets at the air bases where they had landed.

The other air bases where Priština's MiG-21s were deployed on 24 March were under heavy attack. At Podgorica, a MiG-21 was destroyed on the afternoon of 25 March with almost nothing left of it on the ramp. The other one was damaged twice, on 15 and 29 April, and ground crew took it out of the base and saved it from total

Several MiG-21s were destroyed in the Moma Stanojlović Air Depot which was targeted from the first night of Operation Allied Force. (S. Mihajlović)

All of the MiGs of 83rd Regiment which remained outside the underground shelter of Objekt Rudnik were destroyed, as in the case of No. 17208. (M. Špica)

One MiG-21 with a complete AAM load under the wings was destroyed, while two others and a single two-seater were damaged. They were all scattered around the base at the moment of the NATO strike. In an afternoon strike on 11 May, another MiG-21 was destroyed, while another was heavily damaged; both had been spotted on a nearby motorway. In a strike on 17 May between 13:06 and 13:16 another L-17 was destroyed at the road leading from the base to the nearby village of Banovci.

After three of the MiG-21s were destroyed and an additional four damaged in the air attacks between 8 and 17 May, the HQ of 204th Fighter Regiment suggested to the Operations Centre of the Air Defence that the remaining MiGs should be disassembled and taken away from the air base into civilian areas. The regiment's HQ tried to convince the higher authorities about their opinion that the MiGs should be moved out of the base. The MiGs which remained in the base actually represented static targets for the NATO daily strikes. On Friday 21 May, the process of dismantling the aircraft started. However, it was difficult to carry out this task, since NATO carried out intensive daily attacks in this period. On 25, 27 and 28 May, three more MiG-21s were slightly damaged and on 29 May around 02:10, another one was destroyed. On midnight of 7 June, the last of 204th Regiment's MiG-21s was burnt, discovered by NATO aircraft in one of the remaining HASs at Batajnica.[7]

destruction. At Niš, both MiGs were damaged on 16 and 23 April. At Ponikve, one of the two MiG-21s was destroyed on 13 April. Later, as the air strikes continued, other MiGs were damaged: at Niš on 3 May, at Sjenica on 27 May and on 3 June.[6]

An air attack on Batajnica occurred in the early dawn of 8 May, between 03:07 and 03:18, that caused a lot of damage to infrastructure.

Scenes from Objekt Rudnik, 11 June 1999: the ground crew turns No. 17161 toward the entrance. (M. Špica)

One of the last views of the underground Objekt Rudnik with surviving MiGs inside, on 11 June 1999. (M. Špica)

Priština, 11 June 1999: Captain della Croce prior to taking off for Batajnica. (M. Špica)

Preparations for the abandoning of Priština Air Base on 11 June 1999. The three surviving MiG-21UMs seen during the final preparations for taking off for Batajnica. (M. Špica)

As the air campaign stopped around 16:00 on 10 June, the Yugoslav/Serbian forces received the tight schedule for withdrawal from Kosovo to central Serbia. The units of the RV i PVO were ordered to withdraw first from the province. On 11 June, all remaining MiGs from Priština were taken out from the underground Objekt Rudnik. The northern entrance was still blocked following the indirect hit in April. The ground crew had a difficult task to turn around each of the aircraft in the underground gallery, and to take them out of the southern entrance. There were eight MiG-21bis, three MiG-21UMs and a single light piston Utva 75. After a short informative speech by 83rd Regiment's CO, the groups of chosen pilots entered the cockpits and took off for Batajnica.

The first to take off from Priština was a light Utva. Sometime after, eight bis and three two-seaters (each with two pilots onboard) took off flying at low level over the city of Priština, heading to the half-destroyed air base at Batajnica. Radio-silence was ordered. The pilots testified that due to the complete absence of air traffic the silence was total. The feeling was eery.[8]

Priština's MiGs arrived at Batajnica between 16:33 and 17:32. They were welcomed by the half-surprised, half-cautious ground crew of 204th Regiment. On the following day, the other personnel and all of the vehicles of the regiment arrived at Batajnica. After the campaign was over, the remaining damaged MiG-21bis were transported disassembled from Niš and Sjenica air bases where they were initially deployed. At Podgorica, No. 17132 was temporary repaired and brought up to a condition to fly to Batajnica.[9]

During Operation Allied Force, NATO carried out a total of 54 air strikes on 26 days against 83rd Regiment, and 45 air strikes on 204th Regiment on 23 days. Various munitions were used from Mk-82/83/84 iron bombs through to precision bombs and stand-off AGM-130s. Initial strikes were carried out by night, from an altitude of over 7,000 metres and at a distance of 20 kilometres. Later the strikes were

The End of Allied Force

After the stream of political negotiations and shuttle diplomacy reached its peak between 5 and 10 June, the NATO air campaign was halted. The 78-day war was over. President Milosevic accepted that the Yugoslav Army and Serbian Police should abandon Kosovo and turn the security over to the NATO-led Kosovo Force while civil administration should be taken over by the authorities of the UN Mission in Kosovo.

THE ARRIVAL OF 83RD REGIMENT'S SURVIVING FIGHTERS AT BATAJNICA ON THE AFTERNOON OF 11 JUNE 1999

No. 17163 taxies towards the apron. (M. Špica)

Nos.17202 and 17169, just after the engines were shutdown. (M. Špica)

Batajnica's ground crew unload the GSh-23 gun from one of the Priština MiGs. (M. Špica)

Last to arrive were three two-seaters: Nos. 16154, 16155 and 16185. (M. Špica)

carried out during daylight, but the distances were extended to 30-40 kilometres.

Yugoslav fighter aviation carried out a limited number of sorties. Following the decision of the RV i PVO commander, the fleet of the MiG-21 remained grounded. According to the official RV i PVO statistics, 204th Regiment carried 11 interceptor sorties and eight overflights, solely by MiG-29s. 83rd Regiment conducted a total of 28 sorties in deployment from Priština to other air bases, and six more overflights during the air campaign which were regarded as "combat sorties". Detachment 1/353 with reconnaissance MiG-21s also remained inactive.

The authors' research counted a total of 43 MiG-21s lost from an initial 72. Among them 32 were destroyed (23 bis, six UMs and three Ms) and 11 damaged but remained unserviceable (eight bis, three UMs). A total of 29 were used in later years: 20 bis, six UMs, two Rs and a single M.

There was a tendency among the RV i PVO officials to downplay the losses while referring to higher Army or political authorities.[10] They counted only 100% destroyed aircraft or systems, while the damaged assets were not counted as lost. They also did not count the airframes lost at the Air Depot, or they separated fighters from reconnaissance MiGs. Most of the heavily damaged MiG-21s which were not counted as lost were actually in the "beyond repair" category. This was visible in the later period, when several other MiG-21bis or UMs remained stored in Batajnica Air Base or were sent to Moma Stanojlović Air Depot, where they were written off a couple of years later, when the leadership were no longer interested in the facts and figures from the air campaign of 1999.

7
SERBIAN AIR FORCE

Even though the MiG-21s were not used in the combat sorties they were still targeted by NATO whenever they were discovered on the ground and this caused the destruction of nearly half of its fleet. In the summer of 1999, the RV i PVO fighter fleet consisted of the following surviving aircraft, all stationed at Batajnica Air Base:

- 204th Fighter Regiment: 10 MiG-21bis and 3 MiG-21UMs, 4 MiG-29s and 1 MiG-29U. Among all of the available fighters only five MiG-21s were airworthy.
- 83rd Fighter Regiment: 13 MiG-21bis and 3 MiG-21UMs. Those from Priština were in airworthy condition, while others ferried from other air bases awaited repairs.[1]

Upon the orders issued on 14 July 1999 by the HQ of the RV i PVO, many of the units were disbanded. Among them, 83rd Fighter regiment and both of its fighter squadrons: Nos. 123 and 124. 83rd Regiment HQ issued the order on 9 August 1999 for its disbandment. All of their assets, personnel and aircraft were merged into two squadrons of 204th Fighter Regiment.

One of the 20 surviving MiG-21bis was No. 17132. It made the flight from Podgorica a month after Operation Allied Force ended. (MC Odbrana)

Due the expiring of the MiG-29 fleet's resources, MiG-21bis maintained the QRA throughout the 2000s. Here the QRA capabilities are being tested in 2005. (Minga/Cabinet of the Serbian ChoD)

The slow recovery of Moma Stanojlović Air Depot meant that some of the overhauling work was carried out in a rebuilt hangar at Batajnica Air Base. (B. Dimitrijević)

The personnel of 1/353rd Detachment pose near their MiG-21R prior to its withdrawal to the Yugoslav Air Force Museum in 2003. (Nebojša Stefanović Tućko)

Old alliance replaced the short period of intensive belligerence of 1999: a Serbian MiG-21 parked near a USAFE 31st Fighter Wing F-16C during an exchange visit on 22 June 2006. (B. Dimitrijević)

Counting the available MiG-21s, including those that were in the Air Depot, it turns out that 126th Squadron operated 23 (up to 25) MiG-21bis and six MiG-21UMs. Five more MiG-21s were in 1st Detachment of 353rd Reconnaissance Squadron (two Rs, two UMs and one M/LORAP carrier). This makes a total of 36 "functionally serviceable" MiGs in the RV i PVO inventory from mid-1999.

The autumn of 1999 and winter of 2000 were very difficult. 204th Regiment lived in tents in the parts of Batajnica Air Base which were not covered with ruins or cluster bomblets. The lodging problems remained until the middle of 2000. In the same period, the MiG-21bis was reintroduced to the QRA system again after the MiG-29s had serviced this mission since 1988.[2]

The existing MiG-21 pilot community recovered or improved their skills with difficulty. The fleet was reduced almost to a half. The pilots of the whole fighter community from all four pre-war squadrons now gathered in Batajnica totalled just 75. Lack of fuel and the declining resources of the available numbers of MiG-21s of both versions led to flying almost becoming a scarce privilege. It was difficult to enable any kind of improvement to younger and less experienced pilots. Even more, the current commanding officer of the regiment estimated that a "significant portion of the fighter pilots would not have the opportunity to launch or drop air ordnance in their fighter career."[3]

The first post-war live gunnery practice was held in October 2001 at the south Adriatic range off the Montenegrin coast. Eight of the younger pilots that had already carried out overflying missions in the campaign of 1999, made their debut, firing R-60 missiles at SAB-100 illumination bombs acting as the target. Two older pilots carried out firing at targets on the sea surface. A night flying course for the younger pilots followed in 2002, which took much longer than predicted due to lack of fuel. Nine more younger pilots carried out R-60 launches at the south Adriatic range in mid-May 2003.[4]

In this period, the approved annual average of flying hours for MiG-21 pilots was 20 plus three on simulator. It meant that each pilot should perform 30 sorties annually or two to three per month. However, most of the younger pilots did not manage to obtain up to five flying hours annually, thus skipping most of the flying programme. Inherited problems such as only being able to attack from the rear, weaknesses in radar, and poor capabilities of the air surveillance system further limited the weak operational capability of the Yugoslav MiG-21 fleet.[5]

The problem of resources was already obvious in 2001-02. Eighteen of the surviving MiG-21 fleet had been overhauled in the Zmaj Air

Depot prior to 1991, which meant that they were at the limit of serviceability. A few airframes underwent overhauls or different repairs in Moma Stanojlović Air Depot. Estimations dated in 2002, concluded that nearly all resources would expire by the end of 2004.[6]

The available fleet of MiG-21bis started to decline as their resources slowly faded. On 30 May 2003, No. 126 Fighter Squadron operated a total of 20 MiG-21bis, of which seven were serviceable, and five MiG-21UMs, only one of which was serviceable. In this period, 1/353rd Detachment operated two MiG-21Rs, one MiG-21M and one MiG-21UM. At least two more MiG-21bis were in Moma Stanojlović Air Depot at that time waiting for repair, while at least three bis and a UM were ordered to be scrapped.[7]

A serious problem in maintaining the MiGs was the almost complete destruction of Moma Stanojlović Air Depot which had overhauled this type since 1992. Maintenance and overhauling now became serious problems. Some of the overhauling works were carried out in the Batajnica Air Base workshop. A new system of middle-life overhauling was introduced in this period and it enabled much quicker prolongation of the resources for five more years or 500 flying hours. Such practice enabled 204th Regiment to maintain a squadron-size force in full airworthiness and extended the lifespan of the MiG-21s up to 2015.[8]

Sporting Serbian markings, No. 17161 taxies to Batajnica runway. (S. Mihajlović)

Serbian-produced HM-3/23 training-bomblets were introduced for training in bombing missions. (B. Dimitrijević)

In 2005, two of the last MiG-21Rs (26103 and 26105) were retired and towed to the museum at Belgrade International Airport. The 1/353rd Detachment retained a single L-15 (22823) and two piston-engined Utva 75s for flying proficiency. The detachment left 353rd Squadron and in 2006 become 1st Reconnaissance Detachment of 204th Fighter Regiment.

To improve the reconnaissance capability, a new indigenous pod was created to be carried by MiG-21bis. The regular underbelly centreline 490-litre fuel tank (bak) was modified into a reconnaissance pod named "RP-V-D." It was fitted with two Vinten cameras: a panoramic V753A and framing V880, which were controlled by a specialised panel mounted with the pilot's instruments. It was ordered that two of the most recently delivered batch of bis-Ks join the reconnaissance detachment after overhauls in Moma Stanojlović Air Depot. The first – 17407 – arrived at the unit in June 2004, the second – 17409 – during 2006. Even though trials with this indigenous pod were said to be "successful" it did not enter operational service. Both L-17s were returned into their original fighter role and used regularly for QRA and other standard missions.

Final Countdown: MiG-21 in Serbian Air Force Since 2006

After the division of the State Union of the Serbia and Montenegro in May 2006, Serbia remained alone and an independent state. This did not have any influence on the air force or air defence issues, the only result was a change to the aircraft markings which was carried out in 2007-08.

In July 2006, there were 23 MiG-21bis and seven MiG-21UMs available.[9] In October 2006, Mig-21bis No. 17163 was painted in a new camouflage scheme with new Serbian Air Force markings. These were painted on 24 October and officially presented to the public on 15 November 2006, by Serbian President Boris Tadić.[10] The new Serbian roundel was adopted in 2007-08 on all other MiG-21s which remained in operational service.

The reorganisation and downsize of the Serbian Armed Forces was conducted from late 2006 and throughout 2007. In the Air Force, a new system of air bases was introduced and all of the units at one base were now under a single command. On 15 November 2006 at Batajnica, 204th Fighter Regiment and 177th Air Base were merged into 204th Air Base (Avio-baza). Both fighter squadrons Nos. 126 and 127 were merged into a single one, carrying the title: No. 101 Fighter Aviation Squadron. This joint unit operated all available MiG-21s and MiG-29s. 1st Reconnaissance Detachment remained a separate entity in 204th AB. The last remaining MiG-21M (22823) was taken out

Not a common sight: two UZR-60 simulators for R-60 AAMs on the pylons of a Serbian MiG-21bis in 2015. (B. Dimitrijević)

Serbian MiG-21s played the role of their Indian Air Force counterparts in a film co-production. Indian roundels and fin flashes were applied instead of the Serbian ones. (S. Mihajlović)

The last operational sortie of MiG-21bis No. 17409 on the afternoon of the 8 September 2015. (S. Mihajlović)

Table 11: MiG-21s in Yugoslav- and Serbian Service, 1962-2006

Version	delivered	total number	Serial No.	Remark
MiG-21F-13 (L-12)	1962 (5) 1963 (8) 1964 (27) 1966 (1)	41	22501-22541	Used by 204 FR, 117 FR, 83 FR/BR Withdrawn from service in 1981
MiG-21PFM (L-14)	1967 (18) 1968 (18)	36	22701-22736	Used by 204 FR, 117 FR, 128 CC, 129 FS/185, 352 RS, 83 ABr, withdrawn from usage in 1996
MiG-21R (L-14i)	1968 (2) 1969 (6) 1970 (4)	12	26101-26112	Used by 352 RS, 1/353 Det, withdrawn from usage in 2003
MiG-21M (L-15)	1970 (21) 1972 (4)	25	22801-22825	Used by 204 FR, 83 FR, 352 RS, 230 FBS/98 ABr engine changes to MM-16 withdrawn from usage in 1997
MiG-21MF (L-15)	1975	6	22865-22870	Used by 204 FR, 83 FR, modified to L-15M in mid-1980s
MiG-21bis (L-17)	1977 (15) 1978 (18) 1979 (13)	46	17101-112 17125-137 17151-171	Used by 204 FR/ABr, 117 FR/Abr, 83 FR/Abr, VOC, 101 FS/204 AB, withdrawn from use in 2015
MiG-21bis-K (L-17K)	1980 (13) 1981 (12) 1982 (10) 1983 (10)	45	17201-235 17401-409	Used by 204 FR/ABr, 117 FR/Abr, 83 FR/Abr, VOC, 101 FS/204 AB, withdrawn from use in 2015
MiG-21MF (L-15M)	modified in 1984-85 - 4 in 1997 - 2	6	22865-868 22823-824	Modified to carry LORAP Used by 352 RS, 1/353 det withdrawn from usage in 1997 (4) and in 2007 (1)
MiG-21U (NL-12)	1965 (4) 1966 (9) 1967 (5)	18	22901-22918	22901/904 are 66-400 series 22905/918 are 66-600 series Used by all MiG-21 units withdrawn from usage in 1997
MiG-21US (NL-14)	1970	4	22951-22954	Used by all MiG-21 units withdrawn from usage in 1997
MiG-21UM - early (NL-14M)	1971	3	22955-22957	Used by all MiG-21 units engine changed to MM-16 withdrawn from usage in 1997
MiG-21UM (NL-16)	1977 (6) 1978 (6) 1979 (6) 1981 (2) 1982 (4) 1986 (1)	25	16151-160 16171-185	Used by all MiG-21 units, still in use (2020) in 1997 some disarmed to NL-16S (Školski, training) standard, others marked as NL-16b (borbeni, combat)

Two MiG-21UMs remained operational after the official withdrawal of the MiG-21bis fleet. No. 16180, on the right had pylons for AAMs re-fitted to enable the use of R-60 AAMs. (S. Mihajlović)

One of the last sorties of MiG-21UM No. 16178 in spring 2019. (Igor Radić)

from service in 2009, which marked end of the existence of the MiG-21 reconnaissance community.

Problems with prolonged waiting times for repairs and later overhauling of the five remaining Serbian MiG-29s since 2000 have meant that the MiG-21bis remained responsible for the QRA mission during the whole period. Even when overhauled, the MiG-29s soon had other problems with spare parts and tyres which grounded them for another period, leading to the reappearance of the MiG-21 in QRA duties! Even so, after 2010 the MiG-21 fleet started to gradually decline, having MiG-21s drop out of service one-by-one every year.

In the last decade of their usage, MiG-21s appeared at several air shows or open days that were held in Batajnica (2008, 2009, 2012 and 2015) or Niš (2004). They attracted the public with impressive flying capabilities and the thundering roar of their engines. In the later air shows such in 2012 and 2015, the two-seater NL-16 was usually used. Though the MiG-21bis was regarded as obsolete, and its fleet declining year by year, it was still used in combat training and maintaining the QRA until September 2015, when the type was officially retired. The last to conduct a mission was 17409 on the afternoon of 8 September 2015.

The two-seater version remained operational, however, and after the retirement of the MiG-21bis fleet, three of the two-seaters remained in usage. They were all overhauled in recent years and have the possibility of serving for a few more years. Even more, the APU-60 launchers were returned to the wings of No. 16180 enabling it to carry R-60 AAMs and to serve in the QRA system.[11] It was followed in autumn 2016 by No. 16185, making two out of three UMs as stop-gap replacements until further overhauls of the MiG-29s were carried out in the second half of 2018. No. 16185 was painted in the new MiG-29 style scheme.

Three MiG-21UM two-seaters are still in use at the time that this volume is being prepared.

ibliography

Archive sources

Vojni Arhiv Beograd, Srbija (Military Archive Belgrade, Serbia), fund: JNA (Yugoslav People's Army)

Muzej Jugoslovenskog Ratnog Vazduhoplovstva, Beograd, Srbija (Yugoslav Air Force Museum, Belgrade Airport, Serbia),

fund: jedinice RV i PVO (The Units of the RV i PVO)

Uprava Za Organizaciju (Sektora za politiku odbrane Ministartstva odrbane)

-naređenja za organizacijske promene 1949-1999.godine

Yugoslav Air Force RV I PVO Units:

-83. lovački avijacijski puk,with 204th Regiment, at Batajnica Air Base

-117. lovački avijacijski puk, Bihać Air Base

-204. lovački avijacijski puk, Batajnica Air Base

-Vazduhoplovni opitini cenar (Air Test Centre), Batajnica Air Base

Legacy of Colonel Sofronić (in author's possession)

-'Sistem PVO na srednjim i velikim visinama'

-'Studija o PVO teritorije i oružanih snaga SFRJ,'

-'Radari u našim OS'

-Informacija o protivvazdušnoj odbrani naših OS: stanje i mogućnosti, osnovni problemi, pravci daljeg razvoja'

Newspapers and magazines

Aeromagazin, Aeroplan, Glasnik RV i PVO, Front, Krila armije, Narodna armija, Odbramba, Orlovi Plješevice, Politika, YUVAM Bilten.

Literature

The Battle for Slovenia (Cankarjeva Zalozba: Ljubljana. 1991)

Borojević, Danko, Ivić Dragi, Ubović Željko, *Mig 21 legenda hladnog rata* (Stampa: Ruma, 2015)

Čuvari našeg neba (VIZ: Belgrade, 1977)

Della Croce, Roberto, *Priča jednog borbenog pilota* (Stylos Art: Novi Sad, 2013)

Dimitrijević, Bojan., Draganic, Jovica., *Vazdušni rat nad Srbijom 1999. godine* (Institut za savremenu istoriju: Belgrade, 2010)

Dimitrijević, Bojan, *Izviđači Jugoslovenskog ratnog vazduhoplovstva* (Odbrana: Belgrade, 2011)

Dimitrijević, Bojan, *Jugoslovensko ratno vazduhoplovstvo 1942–1992,* III izdanje (Odbrana and Institut za savremenu istoriju: Beograd, 2012)

Dimitrijević, Bojan., Micevski, Milan., *117. lovački puk, 352. izviđačka avijacijska eskadrila, 200. vazduhoplovna baza* (Galaksija: Belgrade – Niš, 2015)

Hamzić, Suad, *Letačke priče* (Tango Six Publishing T6: Belgrade, 2017)

Iskustva iz borbenih dejstava Vojske Jugoslavije u suprotstavljanju agresiji NATO na SR Jugoslaviju, za strategijski nivo (monografija), CŠ VJ, Sektor za ŠONID, (ŠNO: Beograd, October 1999)

Kraljević, Josipa Maras., Vučur, Ilija., *Atentat na Hrvatsku/The Assassination of Croatia* (HDC: Zagreb 2016)

Kronologija Rata, *Hrvatska 1989–1998* (Hrvatski informativni centar: Slovo, Zagreb 1998)

Marijan, Davor, *Smrt oklopne brigade, Prilozi za istraživanje rata za Hrvatsku i Bosnu i Hercegovinu 1990–1992* (Naklada ZORO: Zagreb-Sarajevo, 2002)

Micevski, Milan., Dimitrijević, Bojan., 'Priča o 117. lovačkom puku – Gardijski lovci' *Aeroplan* (Sečovlje, 1/ 1991)

Micevski, Milan., Dimitrijević, Bojan., 'Beg u Austriju', *Aerosvet* (Novi Sad: No 19, Mart 1992)

Micevski, Milan., Dimitrijević, Bojan., *83.lovački puk* (Galaksijanis: Niš, 2016).

Milinković, Colonel Milan, *Operativni zahtevi za dogradnju i modernizaciju lovačke* avijacije, stručni rad (VA, ŠNO GSU 45 klasa, Beograd 2002),

Milikić, Miloš Mido, *Brži od zvuka* (Beograd, 2004).

Milikić, Miloš Mido, *Moji najopasniji letovi* (manuscript, Belgrade 2015)

Oluić, Sasa., Antić, Goran., Dimitrijević, Bojan., *204.lovački avijacijski puk* (VINC: Beograd, 2005)

Osamnaesta klasa VVA, drugo, prerađeno i dopunjeno izdanje (Beograd, 2007)

PRIMORAC Branka, *Perešin, Život i smrt. O Rudolfu, ne samo pilotu...* (Zrinski: Cakovec, 2001)

Radić, Aleksandar, 'Prvi MiG-21 u našem vazduhoplovstvu, Jugosloveni na dva Maha', *Arsenal, specijalni prilog magazina Odrbana* (63, 15 March 2012)

Radić, Aleksandar, 'Druge generacija MiG-21, Čistokrvni presretač', *Arsenal, specijalni prilog magazina Odrbana* (99, 15 March 2015)

Rajtar, Vladimir, *Nebeski ratnici, uspomene hrvatskog pilota* (Vlastita naklada: Zagreb, 1995)

RAZVOJ ORUŽANIH SNAGA SFRJ 1945–1985, knjiga 4, RV i PVO (VIZ: Beograd, 1986)

Sto godina ratnog zrakoplovstva u Hrvatskoj (Despot infinitus: Zagreb, 2012)

Stojić, Jovan, 'Na svetlim tradicijama 254. lovačkog puka', *Glasnik RV i PVO* (Belgarde 3/1978)

Trenadafilovski, Vladimir, *Mikoyan-Gurevich MiG-21 "Fishbed"* (Warpaint series No 91)

Vazduhoplovni opitini centar 1933-2003 ed. Bojan Dimitrijević, (Belgrade, 2003)

VIII klasa ŠAOA, 40 godina na nebu Jugoslavije, 1. novembar 1954–1994 (Beograd, 1994)

DVD

Tajne Željave, Vojni aerodrome Željava kod Bihaća, 2009.

Web page

http://zeljava-lybi.com/forum

CD

204. lovački avijacijski puk, multimedijalna prezentacija (CD), aerodrom Batajnica 2003.

Interviews

Members of the RV i PVO units: Branislav Bajić, Ljubomir Bajić, Tihomir Bašić, Aleksandar Bjelić, Dušan Buban, Nikola Bunčić, Miroslav Čeman, Branislav Dronjak, Ratko Đalović, Rade Đurić, Miša Filka, Vladimir Glišić, Predrag Grandić, Nikola Grujin, Franc Hudomal, Đorđe Ivanov, Zoran Jakovljević, Stanko Janjić, Stevan Janjanin, Savo Jovanović, Luka Kastratović, Gojko Kondić, Dušan Kosanović, Ranko Kovačević, Ranko Krivokapić, Mileta Leković, Rajko Lukić, Vojislav Marković, Božidar Martinović, Milan Menićanin, Bogdan Mihovilović, Momčilo Milenović, Zoran Milićević, Milos Mido Milikić, Milan Milinković, Saša Oluić, Angel Ončevski, Borislav Pejić, Darko Perišić, Slobodan Popović, Nikola Ramljak, Petar Stojković, Vitomir Škrtić, Novica Šušić, Milan Tepšić, Miloš Vukojević, Predrag Vulić, Gvozden Urošević.

Families of: Mate Frulje, Branka Gajovića, Milinka Grbovića, Vlada Kneževića, Vladete Kostića, Slobodana Kusturić, Nikole Maravića, Steve Miletića, Ozrena Radišića, Alekse Ristića, Ilije Zlatića.

Via Tomaž Perme (Ljubljana Slovenia): Vojko Gantar and Alojz Trbovc.

Notes

Chapter 1

1 Vojni arhiv (VA): fund JNA: svežanj 7418/81, 1/5, 'Studija Dinara, 1960'
2 VA: JNA: sv. 8649, 'Izveštaj o radu vojne delegacije na nabavci vojne opreme u SSSR-u, 1961,' Appendix 1: 'Nabavka lovačkih aviona X-2Ma' pp.104-117
3 VA: JNA: sv. 8649, Prilog br. 3: 'Izveštaj o poseti aerodroma Kubinka i prikazu aviona MiG-21F-13', pp.123-137
4 VA: JNA: k. 382, *depeše*
5 VA: JNA: sv. 8649, 'Izveštaj o radu vojne delegacije na nabavci vojne opreme u SSSR-u, 1961'
6 VA: JNA: sv. sv.15407: "Elaborat ROMANIJA, ugovori za nabavke iz SSSR-a 1962-1965"; Muzej Ratnog Vazduhoplovstva (MRV), fund RV i PVO, 'Godišnja istorija za 1962 godinu', p.17
7 Sasa Oluić, Goran Antić, Bojan Dimitrijević, *204.lovački avijacijski puk*, (VINC, Belgrade 2005), pp.49; author's interview with Colonel (ret) Ivan Dubravčić, Zagreb 17 December 2014; Interview with General Angel987Ončevski, New Belgrade, 25 February 2019
8 VA: JNA: k. 386, str.pov. br. 641, 24 September 1962
9 VA: JNA: k. 406. pp.160, 172, 176, 248
10 VA: JNA: k. 387 and 393, 'Prepiska DSNO – vojni izaslanik JNA u Moskvi za 1963. i 1964'
11 Interview with I. Dubravčić
12 VA: JNA: k. 406. p.91
13 MRV; 'Godišnja istorija za 1962 godinu', p.3
14 Interview with A. Ončevski. According to his pilot logbook, Ončevski had his first MiG-21F sortie in Yugoslavia on 8 November 1962
15 Miloš Milikić- Mido, *Moji najopasniji letovi*, Belgrade 2016. (manuscript)
16 MRV: 'Godišnja istorija za 1964 godinu'. Notably, Lockheed TV-2 was the version of the Lockheed T-33A originally manufactured for the US Navy.
17 Milikić, *Moji najopasniji letovi*
18 Interview with A. Ončevski
19 VA: JNA: sv. 3442-1973, p.8
20 VA: JNA: sv. 3442-1973, p.5
21 MRV, 'Godišnja istorija za 1963 godinu', p.21
22 Spomen-soba 204. lovački puk, Batajnica, 'Godišnje istorije 204. vlp' for:1962, 1963, 1964, 1965; *204. lovački avijacijski puk*, multimedijalna prezentacija (CD), aerodrom Batajnica 2003; MRV, 'Godišnja istorija za 1962 godinu', *Operativni sektor;* Ibid, 1964, 'Pozadina i tehnika', Ibid, 1965, 'Pozadina i tehnika (o isporukama R-3S)'; MRV, 'Godisnja istorija za 1965 godinu', p.19; Interview with A. Ončevski; Milikić, *Moji najopasniji letovi*
23 VA: JNA: k. 399, f.1
24 VA: JNA: k. 406
25 VA: JNA: 5387/77, 'Brojno stanje na dan 1.2.1967'
26 VA: JNA: k. 406, p.11; Interview with A. Ončevski
27 VA: JNA: k. 399, telegram No 4969
28 The APA-3MP used for MiG-21F cost 7,500 USD while APA-4 offered for PFM program cost 32,800 USD! VA, k.399, telegram No. 2230
29 VA: JNA: k.406, f.1 telegram No. 6683
30 Interview with A. Ončevski
31 MRV: k. 25, 'Istorija 11.dPVO 1965-76'; Interview with A. Ončevski
32 MRV: k. 25, 'Istorija 11.dPVO 1965-76'
33 MRV: 'Komanda RV i PVO, Analiza VTSl 1969'; *204. lovački avijacijski puk* (CD)
34 MRV: 'Komanda RV i PVO, Pregled stanja VTSl RV i PVO 1970'
35 Bojan Dimitrijević, Milan Micevski, *117. lovački puk, 352.izviđačka avijacijska eskadrila, 200 vazduhoplovna baza* (Galaksija Niš 2015)
36 VA: JNA: svežanj 3442-1973, 'Komanda RV i PVO, izveštaj o borbenoj gotovosti 1964', Ibid, 'Komanda RV i PVO, izveštaj o borbenoj gotovosti 1965'
37 Dimitrijević, Micevski , *117 lovački puk*
38 VA: JNA: svz 3442-1973, 'Komanda RV i PVO, izveštaj o borbenoj gotovosti 1967', p.8
39 Bojan Dimitrijević, *Izviđačka avijacija*, (Odbrana, Belgrade 2011); author's interview with Colonel (ret) Stevo Janjanin, Zemun, 1 November 2002; Suad Hamzić, *Letačke priče*, (Tango six, Belgrade 2017), pp.82-83
40 For details of Bihać AB see the Europe@War volume *Tito's Underground Air Base*. The former Bihać AB is commonly known as 'Željava Air Base' – this is completely inaccurate: no such designation was ever used while the base was still operational (1968-1992). Indeed, Željava is a village in Croatia where only the HQ of 200th Air Base – the administrational element of Bihać AB – was located until 1992 – and from where enthusiasts can nowadays approach Objekt Klek, which was largely demolished during the withdrawal of the RV i PVO in May 1992. The word "objekt" can be loosely translated in this context as "facility".
41 MRV: k.31, 'Istorija 117. vp za 1968', Dimitrijević-Micevski, *117. lovački puk*, pp.72–75; Personal notes of Colonel Ilija Zlatić, which the authors obtained from his family, January 2015.
42 MRV: k.25, 'Istorija 11.dPVO 1965-76' (1968)
43 Author's interview with Colonel (ret) Ivan Katić, Radio-far, September 1993
44 Interview with I. Dubravčić; MJRV, k.25, 'Istorija 11.dPVO 1965-76' (1968)
45 VA: JNA: sv. 5387/77, 'Brojno stanje na dan 1.11.1968'
46 MRV: 'Komanda RV i PVO, Analiza VTSl 1969'; Ibid, 1971; Hamzić, *Letačke priče*
47 MRV: k. 112a, (variae dated with 1969)
48 MRV: k. 27: '5. VaK , Godišnje istorije 1972, 1973 and 1974'; author's interview with Colonel (ret) Novica Sušić New Belgrade, 22 October 2002; Interview with S. Janjanin

Chapter 2

1 MRV: 'Komanda RV i PVO, Pregled stanja VTSl RV i PVO 1970'
2 MRV: 'Komanda RV i PVO, Pregled stanja VTSl RV i PVO 1970'; MRV: k.25, 'Istorija 11.dPVO 1965-76'; Oluić, Antić, Dimitrijević, *204. lovački puk*, p.61; author's interview with General (ret) Živan Mirčetić Belgrade, 16 October 2002
3 MRV: 'Komanda RV i PVO, Pregled stanja VTSl RV i PVO 1970', p.11
4 Dimitrijević-Micevski, *117. lovački puk*, pp.78-79
5 MRV: 'Komanda RV i PVO, Pregled stanja VTSl RV i PVO 1970'
6 MRV, 'Godišnja analiza rada VTS i VT obezbeđenje manevra SLOBODA 71'; MRV: k.25, 'Istorija 11.dPVO 1965-76' (1971); interview with Ž. Mirčetić
7 MRV: k.25, 'Istorija 11.dPVO 1965-76' (1974)
8 Author interview with Colonel (ret.) Ranko Kovačević, December 2014
9 MRV, RV i PVO: k.27, 'Istorija 5. VaK, 1964-1972'
10 Hamzić, *Letačke priče*, pp.149-150; author's interview with Colonel (ret) Novica Šušić, Novi Beograd, 22 October 2002
11 Interview with N. Šušić; Interview with S. Janjanin; Hamzić, *Letačke priče*, pp.191-202
12 Author interview with Colonel (ret.) Gojko Kondić Belgrade, February 2015
13 Interview with N. Šušić; Interview with S. Janjanin; Hamzić, *Letačke priče*, p.42-151
14 Hamzić, *Letačke priče*, pp.114-117
15 Hamzić, *Letačke priče*, pp.230-232; Interview with R. Kovačević
16 Milan Micevski, Bojan Dimitrijević, *83. lovački puk*, (Galaksijanis, Niš 2016), pp.69–70
17 Micevski, Dimitrijević, *83. lovački puk*, p.77
18 M. Micevski, B. Dimitrijević, *83. lovački puk*, p.77
19 Interview with A. Ončevski
20 Micevski, Dimitrijević, *83. lovački puk*, pp.81-90
21 Rajica Bošković, *Nebo na dlanu, Vek vazdušnog osmatranja, javljanja i navodjenja*, (Author's Edition, Beograd 2017), pp.235-245
22 Bošković, *Nebo na dlanu*, pp.235-245; M. Micevski, B. Dimitrijević, *83. lovački puk*, pp.91–92
23 Author interview with General (ret.) Radomir Gordić, Belgrade 2016
24 Naredba Komande RV i PVO, 13 December 1977
25 VA: JNA: k. 406, 45; MJRV, 'Komanda RV i PVO, analiza rada VTSl za III tromesečje 1967'
26 MJRV: 'Komanda RV i PVO, analiza rada VTSl za 1969.'
27 According to the duration of stay in the air depot, registered on lists of the airplanes sent to Zmaj Air Depot for overhaul.
28 Interview with I. Dubravčić
29 Interview with Vitomir Škrtić, Zagreb 10 April 2019
30 MRV: 'Komanda RV i PVO, analiza rada VTSl za 1969.'
31 MRV: 'Komanda RV i PVO, analiza rada VTSl za 1969.' Ibid, 'Pregled stanja VTSl RV i PVO 1970'

Chapter 3

1 'Sistem PVO na srednjim i velikim visinama', offical study of Department for PVO, (legacy of Colonel Sofronić), pp.14-15
2 Interview with Ž. Mirčetić; 204. lovački puk 'Istorija 204.lap za 1977. godinu'
3 MRV: k. 25, 'Istorija 11.dPVO 1977', p.25
4 Spomen-soba 204. lovački puk: 'Godišnje istorije' 1977 to 1982; *204. lovački avijacijski puk,* (CD); Oluić, Antić, Dimitrijević, *204. lovački puk*, pp.69–70
5 Spomen-soba 204. lovački puk: 'Godišnja istorija 1978'; MRV, k.25, 'Istorija 11.dPVO' 1978, p.18
6 MRV, k.25, 'Istorija 11.dPVO' 1978, 1979, 1980
7 SSNO, GŠ OS SFRJ, ZNGŠ OS SFRJ za RV i PVO, Uprava PVO, 'Studija o PVO teritorije i oružanih snaga SFRJ,' NO-DT primerak br. 8, 1989. p.70
8 Spomen-soba 204. lovački puk: 'Godišnja istorija 1983'
9 MJRV, k. 31, istorija 117 .lap (1978, 1979, 1980) B. Dimitrijević, M. Micevski, *117. lovački puk*, pp.106–107
10 MRV, RV i PVO: k.31, 'Istorija 117. vp 1980'
11 Oluić, Antić, Dimitrijević, *204. lovački puk*, pp.71-72; author's interview with Colonel (ret) Božidar Martinović, Novi Beograd 21 July, 20 and 22 August 2016
12 MRV: k.25, 'Istorija 11.dPVO' 1977
13 MRV: k.25, 'Istorija 11.dPVO' 1981; Micevski, Dimitrijević, *83. lovački puk*, pp.96–97
14 Hamzić, *Letačke priče*, p.281
15 Komanda 128.centra za preobuku pilota na supersoničnim avionima: 'Program obuke na avonu NL i L-14 pilota oružanih snaga Palestinske oslobodilačke organizacije'; author's interview with Lt.Colonel (ret) Momčilo Milenović and Major (ret) Savo Jovanović, 31 May 2019

16 MRV:k.25, 'Istorija 11.dPVO' 1982;Spomen-soba 204. lovački puk,'Godišna istorija 1982'
17 Naredba SSNO, 2157–1, od 24. decembra 9, 17 and 24 March 2019 1980
18 MRV: k. 36, 185. lbap (Š), (regimental histories between 1983 and 1990)
19 *Informacija o protivvazdušnoj odbrani naših OS (stanje i mogućnosti, osnovni problemi, pravci daljeg razvoja)* s/pov, datirano: 12. april 1985 (legacy of Colonel Sofronić), p.2
20 Author's interview with Franc Hudomal, Novi Beograd, spring 2016
21 MRV: k.31, '117.lap' (1981–1983); Ibid, k.31V, '117.lap' (1984–1985)
22 MRV: k.27, '5. VaK', 1978
23 MRV: k.27, '5. VaK', 1979
24 MRV: k.25, 'Istorija 11.dPVO' 1983
25 Spomen-soba 204. lovački puk: 'Godišnja istorija 1983'
26 Author's interview with Colonel (ret) Zoran Milićević, Novi Beograd 2 April 2016; author's interview with Lieutenant Colonel (ret) Nikola Ramljak Novi Beograd, February 2016
27 MRV: k.31, 'Istorija 117. vp za 1983'
28 Dimitrijević, *Izviđačka avijacija*, pp.114-116
29 Interview with S. Janjanin
30 MRV: k. 87, '352. izviđačka avijacijska eskadrila' (1987, 1988, 1989)
31 Naredba SSNO 28 February 1986, 'Plan *Jedinstvo*' ; Naredba 1514–1/87 17 Septembar 1987
32 Elaborate 'Radari u našim OS', pomoćnik NGŠ JNA za elektroniku i veze, April 1987. (Sofronić), pp.10–11
33 'Studija o PVO teritorije i oružanih snaga SFRJ', pp.69-70
34 Spomen-soba 204. lovački puk: 'Godišnje istorije' 1986,1987; Oluić, Antić, Dimitrijević, *204. lovački puk*, pp.79–81
35 'Studija o PVO teritorije i oružanih snaga SFRJ'
36 As an illustration, a list dated 1 November 1987, showed that the RV i PVO operated following numbers of AAMs; 1,604 K-13 (R-3Ss), 1,370 R-3Ms, 1,093 R-3Rs, 546 R-60s and 157 K-51s. Colonel Dušan Bosković, *Radna beležnica*, p. 187, personal notebook given to the authors on 18 August 2016.
37 Aleksandar Radić, *Prvi MiG-21 u našem vazduhoplovstvu, Jugosloveni na dva Maha*, Arsenal, (specijalni prilog magazina Odrbana, 63, Belgrade, 15 March 2012). p.63; Aleksandar Radić, *Druge generacija MiG-21, Čistokrvni presretač*, Arsenal, (specijalni prilog magazina Odrbana, 99, Belgrade 15 March 2015), p.29
38 VA, k.399, depeše
39 Vazduhoplovni opitni centar (Air Test Centre, VOC): 'Izveštaji o ispitivanjima vazduhoplova od 1980'. *Vazduhoplovni opitini centar 1933-2003* ed. Bojan Dimitrijević, (Belgrade, 2003).
40 VOC, 'Izveštaji o ispitivanjima vazduhoplova 1992-1999'
41 MRV: k.26, '3 korpus RV i PVO', 1989
42 Naredba SSNO drzavna tajna 432-1, 5 February 1990 and 432-12, 23 July 1990: 'Plan Jedinstvo2/3'; author's interview with Lt. Colonel Darko Perišić, Belgrade, 28 September 2002

Chapter 4

1 B. Dimitrijević, M. Micevski, *117. lovački puk*, p.134
2 Naredba: Komanda RV i PVO, s/pov. br. 03/354–397, 25 June 1991
3 B. Dimitrijević, M. Micevski, *117. lovački puk*, p.137
4 'Svedočanstvo sa kriznih mesta, (ekskluzivan Politikin izveštaj sa lica mesta oružanih akcija)', *Politika*, (28 June 1991), pp.1–2; *The Battle for Slovenia*, (Cankarjeva zalozba Ljubljana 1991) pp.28–34; Dimitrijević, Micevski, *117. lovački puk*, p.137
5 'Saopštenje Komande RV i PVO (10,30) TO Slovenije napada helikoptere sa hranom,' *Borba* (3 July 1991), 2; 'Lažni mir pravi–rat', *Novosti*, (3 July 1991), p.5; 'Na vročih položajih', *Revija Odbramba*, 6/1998, pp.14–15; *The Battle for Slovenia*, pp.78–89
6 Dimitrijević, Micevski, *117. lovački puk*, p.137
7 Dr Vaso Predojević, *U prodcjepu* (DanGraf , Belgrade 1997), passim; Janez Janša, *Premiki, Nastajanje in obramba slovenske države 1988–1992* (Mladinska knjiga Ljubljana 1992). On 12 August the remaining assets of 82nd Aviation Brigade, now as 238th Fighter-Bomber Squadron, moved to Banja Luka Air Base and became part of 117th Regiment. This now became 117th Aviation Brigade with three MiG squadrons and one Orao/Jastreb strike squadron. The brigade became responsible for most of the combat sorties over Croatia which would be carried out in the continuation of the war during the autumn and winter 1991-92. Naredba SSNO, 2284–1, 30 August 1991; 'Ponovo krila nad Zalužanima', *Narodna armija*, (14 August 1991), p.4; 'Novi vojni aerodrom RV i PVO kod Banja Luke,' *Politika* (13 August 1991)
8 Author's interview with Lt Colonel Saša Oluić, Belgrade, 2 April 2016; author's interview with Miroslav Čeman , Belgrade, 10 June 2016
9 'Nove pojedinosti o krijumčarenju oružja ugandskim avionom', *Narodna armija*, (11 September 1991), pp.10–11. *„Mig" u „zenge"Narodna armija*, (11 September 1991), pp.27; *Kronologija rata, Hrvatska 1989 –1998*, (Hrvatski informativni centar; Slovo, Zagreb 1998), p.89. See also the Europe@War companion volume *The Yugoslav Air Force in the Battles for Slovenia, Croatia, and Bosnian and Herzegovina Volume 1.*
10 *Kronologija rata, Hrvatska 1989-1998*, p.92
11 Naredba DSNO, 2944–1, 30 October 1991; author's interview with W/O (ret) Miroslav Špica 19 April 2019
12 'Lekcija otimačima i švercerima oružja,' *Narodna armija*, (18 September 1991), p.7; *Sto godina ratnog zrakoplovstva u Hrvatskoj*, (Despot infinitus, Zagrerb 2012). pp.357
13 Author's interview with Lt Colonel (ret) Branislav Dronjak, 11 February 2015
14 Author's interview with Colonel (ret) Dušan Buban 9 January 2015; Dimitrijević, Micevski, *117. lovački puk*, pp.141
15 Roberto della Croce, *Priča jednog borbenog pilota*, (Stylos Art Novi Sad, 2013), pp.19-40
16 Dimitrijević, Micevski, *117. lovački puk*, p.192
17 'Onesposobljen RTV odašiljač na Sljemenu,' *Politika*, (5 October 1991), p.9; 'Avioni i artiljerija JNA zaustavili kolonu tenkova Hrvatske,' *Politika*, (5 October 1991), p.9
18 'Avijacija nije bombardovala Banske dvore' and 'Banski dvori nisu pogođeni raketom iz aviona' *Politika*, (9 October 1991), p.5; 'Scenario s greškama' and 'Izmišljeni napadi na Zagreb', Narodna armija, (21 September 1991), p.7; Josipa Maras Kraljević, Ilija Vučur, *Atentat na Hrvatsku/The Assassination of Croatia*, (HDC Zagreb 2016), pp.19–21
19 Branka Primorac, *Perešin, Život i smrt. O Rudolfu, ne samo pilotu…*, (Zrinski; Cakovec, 2001), pp.104–120; *Kronologija rata, Hrvatska 1989–1998*, p.109; Milan Micevski i Bojan Dimitrijević, 'Beg u Austriju', *Aerosvet* (Novi Sad; No 19, Mart 1992), pp.40–41
20 'Piloti dejstvovali efikasno', *Narodna armija*, (9 November 1991), p.7; 'Snažna podrška Kopnenoj vojsci' and 'Razorena dva skladišta', *Narodna armija*, (13 Novembar 1991), p.6; 'Duga lista uspešnih dejstava', *Narodna armija*, (16 November 1991), p.6
21 Interview with S. Janjanin; Document: Policijska ispostava Ravna Gora 'Službena zabilješka', 10.11.1991, (courtesy by Dean Čanić)
22 della Croce, *Priča jednog borbenog pilota* , pp.19–40; Dimitrijević, Micevski, *117. lovački puk*, p.154.
23 Komanda 204.lap: st.pov 03/3-38 16 March 1992: 'Ratni put 204.lap-a od 26.06 do 31.12.1991.g'
24 Komanda 3.korpusa RV i PVO: pov br 25/45-2, 03 04 1992: 'Istorija 3.ko RV i PVO 1991', appendix: 7
25 Micevski, Dimitrijević, *83. lovački puk*, pp.135-136
26 Interview with V. Škrtić
27 'Mesečni, izveštaj vazduhoplovnotehničke službe o vazduhoplovima', VOB-110, str. poverljivo (courtesy Dragan Kolundžić)
28 'Avioni bez oznake preleću embargo', *Politika*, (24 November 1993), p.9; Aleksandar Radić, *Avion MiG-23 sa našim oznakama, Ratom oblikovana sudbina*, Arsenal, (specijalni prilog magazina Odbrana, br. 57), pp.30–32
29 Dimitrijević, Micevski, *117. lovački puk*, pp.145–146
30 Komanda 204. lap: s/pov, 03/3–38, 16, March 1992
31 Komanda 204. lap: s/pov, 03/3–38, 16, March 1992
32 MRV: 'Analiza VTSl za 1991. godinu'
33 Savezni sekretarijat za saobraćaj i veze: br. VII-5, 1/6, 18 January 1992, Komisija za ispitivanje uzroka udesa vazduhoplova; Komisijski izveštaj, 1–19 (courtesy by Živan Mirčetić, April 2003)
34 *Kronologija rata, Hrvatska 1989–1998*, pp.128, 134 and 137
35 *Sto godina ratnog zrakoplovstva u Hrvatskoj*, pp.362–363; Dimitrijević, Micevski, *117. lovački puk*, p.157
36 Author's interview with D. Buban and B. Dronjak; Dimitrijević, Micevski, *117. lovački puk*, pp.150–151
37 *Orlovi Plješevice*, issues for 1992; Dimitrijević, Micevski, *117. lovački puk*, p.151
38 Author's interview with D. Buban and B. Dronjak; Dimitrijević, Micevski, *117. lovački puk*, p.151
39 Spomen-soba 204. lovački puk: 'Istorija 204. lap/abr za 1992. godinu'
40 Spomen-soba 204. lovački puk: 'Istorija 204. lap/abr za 1992. godinu'; Interview with D. Perišić
41 Davor Marijan, *Smrt oklopne brigade, Prilozi za istraživanje rata za Hrvatsku i Bosnu i Hercegovinu 1990–1992*, (Naklada ZORO, Zagreb-Sarajevo 2002). pp.83–136
42 Author's interview with Lt Colonel (ret) Predrag Grandić, Belgrade, March 2015
43 Dimitrijević, Micevski, *117. lovački puk*, p.152; della Croce, *Priča jednog borbenog pilota*, pp.19–40
44 MRV: k.25-I, 'Istorija VaK' 1992, p.36; Oluić, Antić, Dimitrijević, *204.lovački avijacijski puk*, pp.91–92
45 Author's interview with Colonel (ret) Milan Milinković autoru, Zemun Polje, 4 June 2016; Micevski, Dimitrijević, *83. lovački puk*, p.136
46 MRV: k.25-I, 'Istorija VaK' 1992, p.36; Interview with M. Špica
47 Author's interview with Z. Milićević; Micevski, Dimitrijević, *83. lovački puk*, pp.137–139
48 'Dva pilota prebegla sa migovima 21 u Hrvatsku,' *Politika*, (17 May 1992), p.1; 'Jedan sleteo u Zagreb drugi u Split', *Politika*, (18 May 1992), p.5; 'Smenjene odgovorne starešine aerodorma u Užicu,' *Politika*, (19 May1992)
49 MRV: k.25-I, 'Istorija VaK' 1992, pp.37-38

Chapter 5

1 MRV: k.25-I, 'Istorija VaK' 1992'
2 Spomen-soba 204. lovački puk: 'Istorija 204. lap/abr za 1992. godinu'; MRV, k.25-I, 'Istorija VaK' 1992, pp.14-17, 21-25 and 34
3 Micevski, Dimitrijević, *83. lovački puk*, p.139-141

continued on page 72

ACKNOWLEDGEMENTS

The authors wish to express their gratitude to all the kind colleagues who helped us with unpublished photographs from their respective institutions: Radovan Cukić (Museum of Yugoslavia, Belgrade), Amir Obhođaš and Hrvoje Gržina (Croatian State Archive, Zagreb), Zoran Milovanović (MC Odbrana), Lt. Colonel Predrag Grandic (Yugoslav Air Force Museum), Colonel (ret.) Vojislav Marković (83rd Fighter Regiment), Lieutenant-Colonel Ivan Sitar (204th Fighter Regiment) and Lieutenant-Colonel Saša Tirnanić (Air Test Centre). Especially, to the widely known air photographer Milorad Ristić who provided us his photos taken in Bihać during 1991. Also, to Warrant Officer Miroslav Špica who had taken valuable photos during the air campaign of 1999 at Priština Air Base and to Warrant Officer Slobodan Mihajlović "Mixelotti".

To General (Ret.) Angel Ončevski, Colonels (ret.) Suad Hamzić, Stevo Janjanin, Novica Šušić, Borislav Pejić, Bogdan Mihovilović, Ranko Kovačević, Momčilo Milenović, Miloš Vukojčić, Siniša Šijačić, Zoran Milićević, Stanko Janjić and families of Generals Nikola Maravić and Branko Gajević, Colonel Franc Ficko.

Also, to those who provided us with other valuable information and documents: General Jovica Draganić, Colonel Milan Milinković, Lieutenant-Colonel Saša Oluić and Major Miroslav Čeman (RV i PVO), Dragan Kolundžić (Pančevo, Serbia), Vladimir Šumanovac, Dean Čanić (Zagreb, Croatia) and Dr Nikica Baric and Dr Davor Marijan from History Institute in Zagreb, Croatia.

Finally, to Tom Cooper who suggested to the authors this unique subject related to the Yugoslav Cold War aviation history.

ABOUT THE AUTHORS

Bojan Dimitrijevic is the Deputy Director of the Institute for Contemporary History in Belgrade, Serbia. Educated at the Universities of Belgrade and Novi Sad, CEU Budapest and University of Bradford, he was custodian of the Yugoslav Aviation Museum before, serving as advisor to the Minister to the Serbian MoD, Minister of Foreign Affairs, the President of Serbia, and as Assistant to the Minister of Defence. His has researched the military history of Yugoslavia, and the Balkans in the Second World War, Cold War, and conflicts since, he has published over 50 books and 100 scientific articles in Serbia and abroad. This is his fourth book for Helion's @War series

Milan Micevski is an entrepreneur from Belgrade, he has been exploring archives and collecting data on aviation history for more than three decades. He is considered one of the leading experts in the field of the Yugoslav Air Force as well as Soviet Aviation in former Yugoslavia. With Bojan Dimitrijević, he has so far co-authored seven books on different aspects of the Yugoslav Air Force history, and has worked with other authors two other volumes. This is his second instalment for Helion's @War series.

continued from page 71

4 Micevski, Dimitrijević, *83. lovački puk*, p.139
5 Author's interview with Colonel (ret) Zoran Jakovljevic, Belgrade, December 2015
6 MRV: k.25b, 'Istorija Korpusa PVO' 1994
7 MRV: k.25-I, 'Istorija Vazduhoplovnog korpusa' 1994; MRV: k.33 'Istorija 98.avijacijske brigade', 1994
8 MRV: k.25b, 'Istorija Korpusa PVO' 1995 and 1996
9 VOC, 'Izveštaji o ispitivanjima vazduhoplova 1992-1999'
10 MRV: k.27B, 'Istorija Vazduhoplovnog korpusa' 1997

Chapter 6

1 della Croce, *Priča jednog borbenog pilota* , pp.30-40
2 Spomen-soba 204. lovački puk: 'Godišnja istorija' 1998
3 Colonel Milan Milinković, *Operativni zahtevi za dogradnju i modernizaciju lovačke* avijacije, stručni rad, (VA, ŠNO GSU 45 klasa, Beograd 2002), p.41
4 Spomen-soba 204. lovački puk: 'Godišnja istorija' 1999
5 Spomen-soba 204. lovački puk: 'Godišnja istorija' 1999
6 Micevski, Dimitrijević, *83. lovački puk*, pp.157-158
7 Spomen-soba 204. lovački puk: 'Godišnja istorija' 1999
8 della Croce, *Priča jednog borbenog pilota* , pp.50; Micevski, Dimitrijević, *83. lovački puk*, pp.160-162
9 Milinković, *Operativni zahtevi za dogradnju i modernizaciju*, pp.42
10 *Iskustva iz borbenih dejstava Vojske Jugoslavije u suprotstavljanju agresiji NATO na SR Jugoslaviju, za strategijski nivo* (monografija), ČŠ VJ, Sektor za ŠONID, (ŠNO, Belgrade, October 1999)

Chapter 7

1 'Mesečni, izveštaj vazduhoplovnotehničke službe o vazduhoplovima', VOB-110, July 1999
2 Spomen-soba 204. lovački puk: 'Godišnja istorija' 2000
3 Spomen-soba 204. lovački puk: 'Godišnja istorija' 2000; Milinković, *Operativni zahtevi za dogradnju i modernizaciju*
4 Spomen-soba 204. lovački puk: 'Godišnja istorija' 2000; Oluić, Antić, Dimitrijević, *204. lovački puk*
5 Milinković, *Operativni zahtevi za dogradnju i modernizaciju*, pp.56-59
6 Milinković, *Operativni zahtevi za dogradnju i modernizaciju*, pp.44
7 Komanda 204.lap: VTO, str.pov br.40/28-21, 30 May 2003; VZ Moma Stanojlović: 'Pregled vazduhoplova na opravci u VZ Moma Stanojlović za mesec maj 2003 godine'
8 Komanda V i PVO: 'Vremenski rokovi rada aviona L-17, NL -16', 4 and 13 July 2006, courtesy of General Jovica Draganić
9 Komanda V i PVO: 'Vremenski rokovi rada aviona L-17, NL -16'
10 One of the authors of this volume – as advisor the President – was the designer of the adopted roundel.
11 Živojin Banković, Prva domaća modifikacija lovačkog aviona u Srbiji: RV i PVO integrisalo rakete R-60 na dvosede MiG-21UM < https://tangosix.rs/2016/12/04/mig-21um-sa-r-60/ > (accessed 15 March 2019)